Math Mammoth Multiplication 2

By Maria Miller

ISBN 978-1-954358-71-3

2020 EDITION

Contents

Introduction

Math Mammoth Multiplication 2 is a worktext focusing on multi-digit multiplication and related topics. It best suits fourth grade mathematics.

The first lessons briefly review the concept of multiplication and the multiplication tables. Next, students encounter equations in disguise — presented with shapes on both sides of a pan balance — in the lesson *Scales Puzzles*. This lesson is intended to be fun and motivational.

Then, the focus shifts to multi-digit multiplication (also called multiplication algorithm or multiplying in columns). We start out by learning to multiply numbers by multiples of ten and hundred (for example, 20×4 or 500×6). After this is mastered, students learn the very important concept of **multiplying in parts**, or partial products. This means that, for example, we multiply 4×63 in two parts: first we multiply $4 \times 60 = 240$ and $4 \times 3 = 12$, and lastly the results are added: $240 + 12 = 252$.

This principle underlies all other multiplication algorithms, so it is important to master. We don't want children to "blindly" memorize the multiplication algorithm without understanding what is going on with it. The partial products algorithm (multiplying in parts) also ties in with an area model, and it is very important that students see the connection between this visual model and the procedure.

The book contains two lessons about multiplying in columns the "easy way". This "easy way" is a simplified form of the traditional multiplication algorithm, based on partial products. You may skip these two lessons at your discretion. The method taught in those lessons is most useful for students who may have trouble with the traditional form of the algorithm. This method is also helpful in cementing the student's understanding of the partial products method.

The traditional, or standard, form of multiplication algorithm is taught next, and is hopefully fairly easy, with the partial products as a foundation.

Students also study estimation, the order of operations, and multiplying with money. There are numerous word problems in all of the lessons. Students are encouraged to write number sentences for the word problems—essentially learning to show their work and their thinking process.

The lesson "*So Many of the Same Thing*" has to do with proportional reasoning. The idea is really simple, and prepares students for learning ratios and proportions in middle school.

The last major topic in the book is multiplying two-digit numbers by two-digit numbers. Again, we first study partial products and tie that in with an area model. The lesson *Multiplying in Parts: Another Way* is optional. Lastly, the book teaches the standard algorithm for two-digit by two-digit multiplication. Students will practice multiplication with more digits in the book *Math Mammoth Multiplication & Division 3*.

I wish you success in teaching math!

Maria Miller, the author

Helpful Resources on the Internet

We have compiled a list of external Internet resources that match the topics in this book. This list of links includes web pages that offer:

- **online practice** for concepts;
- online **games**, or occasionally, printable games;
- **animations** and interactive **illustrations** of math concepts;
- **articles** that teach a math concept.

We heartily recommend you take a look at the list. Many of our customers love using these resources to supplement the bookwork. You can use the resources as you see fit for extra practice, to illustrate a concept better, and even just for some fun. Enjoy!

https://l.mathmammoth.com/blue/multiplication2

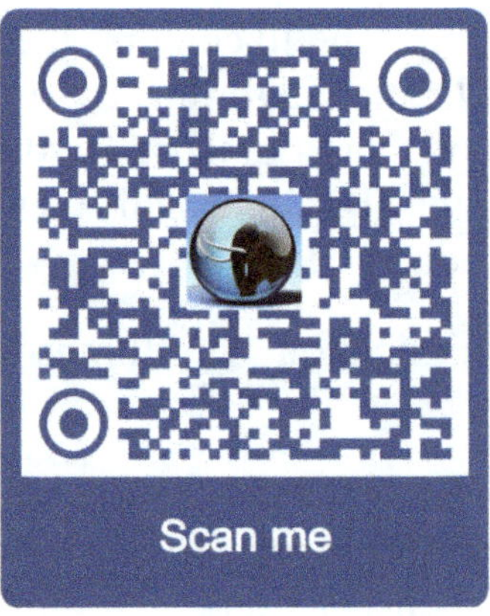

Understanding Multiplication

- Multiplication has to do with many groups of the same size: 3×5 means three groups of 5. You can find the total by adding: $3 \times 5 = 5 + 5 + 5 = 15$.
- Multiplying by 1 means you have just one group: $1 \times 17 = 17$.
- Multiplying by 0 means "no groups": $0 \times 82 = 0$
- The order in which you multiply does not matter: 3×6 and 6×3 are both 18.

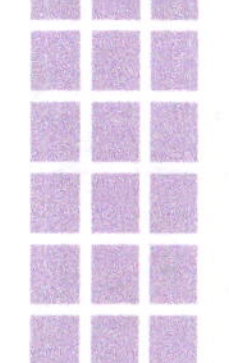

3 groups of 6 *or* 6 groups of 3.

1. Write the additions as multiplications, or vice versa. Solve.

a. $2 + 2 + 2 + 2 =$ ____ × ____ = _____	**b.** $8 + 8 + 8 =$ ____ × _____ = _______
$20 + 20 + 20 + 20 =$ ____ × ____ = _____	$80 + 80 + 80 =$ ____ × ____ = ______
c. ________________________________ $= 4 \times 500 =$ _________	
________________________________ $= 3 \times 120 =$ _________	

2. Write two multiplications.

a. _____ × _____ = ______ _____ × _____ = ______ 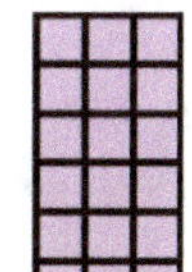	**b.** _____ × _____ = ______ _____ × _____ = ______ 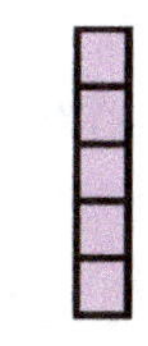

3. Solve.

a. $8 \times 2 =$ _____	**b.** $3 \times 5 =$ _____	**c.** $2 \times 8 =$ _____	**d.** $3 \times 10 =$ _____
$8 \times 0 \times 7 =$ _____	$1 \times 2 \times 5 =$ _____	$2 \times 2 \times 2 =$ _____	$3 \times 3 \times 3 =$ _____

4. Find the products. You can often use addition.

a.	b.	c.	d.
$2 \times 24 =$ _____	$2 \times 150 =$ _______	$4 \times 1{,}000 =$ ________	$2 \times 34 =$ ______
$14 \times 0 =$ ____	$3 \times 2{,}000 =$ _______	$5 \times 200 =$ ________	$3 \times 21 =$ ______
$16 \times 1 =$ _____	$4 \times 3{,}000 =$ ________	$3 \times 211 =$ _________	$4 \times 50 =$ ______

Multiplication terms

The numbers being multiplied are **factors**.
The result is called a **product**.

There may be more than two factors. For example, in $4 \times 5 \times 2 = 40$, the numbers 4, 5, and 2 are all factors.

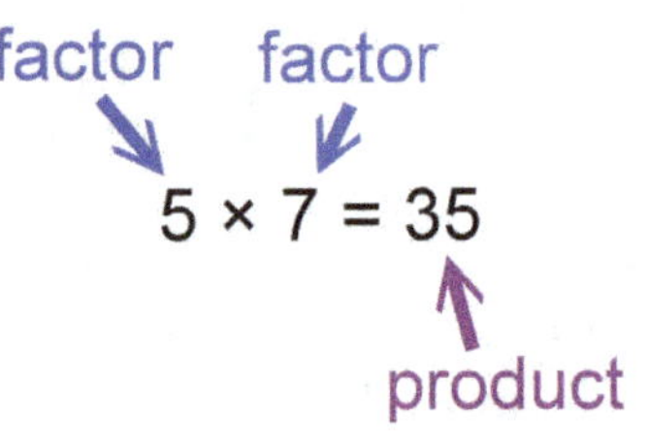

5. Find the unknown factors.

a. ______ × 2 × 2 = 24	b. 3 × ________ = 600	c. 500 × ____ = 1,500
______ × 9 × 2 = 0	4 × ________ = 1,000	10 × _____ = 810

6. Fill in.

a. Write the the terms. 2 × 23 = 46 (↑ ↑ ↑) ____________ ____________	b. Write a multiplication problem with factors 4 and 8.
c. What happens if one of the factors is zero? The _____________________ is ________.	
d. In one multiplication problem, two factors are 2 and 6. The product is 60. What is the third factor?	

7. Write a number sentence for each of these problems. Use several operations in it.

Problem:	Number sentence:
a. Mom had three dozen eggs in cartons and five in a bowl. How many eggs did she have in all?	3 × 12 + 5 =
b. Jack bought six packages of magazines. Each had 10 magazines. He opened one package and gave three magazines to his friend. How many magazines does Jack have left?	
c. Anna had seven boxes. Into four of the boxes, she put 10 crayons each, and into three boxes she put only 6. How many crayons did she use?	
d. Ernest bought three books for $11 each and paid with $50. What was his change?	
e. How many wheels do five tricycles and seven bikes have in total?	

Example. A simple hat costs \$6. Another, fancier hat, costs \$18.
How many times more expensive is the fancier hat?

It asks "how many times", so that is our unknown (_?_). We write a multiplication:

$$\underline{\ ?\ } \times 6 = \$18$$

It is easy to see the answer is three times, or _?_ = 3.

8. Solve the problems. Write a multiplication with an unknown (_?_ or y) for each problem. What is the unknown in each problem? It is what the problem asks for or what you do not know.
(Note also: we are not using x as an unknown, as it could be confused with the multiplication sign ?.)

a. Each child has 10 toes. How many toes would seven children have? _____ × _____ = _?_ _?_ = _________	**b.** If each cow has four feet, how many cows are there if there is a total of 24 feet? _?_ × _____ = _______ _?_ = _________
c. One bicycle has two wheels. _?_ bicycles have 18 wheels. _____ × _____ = _______ _?_ = _________	**d.** One car has 4 wheels. So, y cars have 36 wheels. _____ × _____ = _______ y = _________
e. How many people would you need to have a total of 150 fingers? _____ × _____ = _______ y = _________	**f.** How many dozen eggs would be 60 eggs? _____ × _____ = _______ y = _________
g. Carl owns 20 children's books. Emma owns four times as many children's books. How many children's books does Emma own? _____ × _____ = _______ y = _________	**h.** You can fit 7 people in a van. How many such vans do you need to take 35 people to the beach? _____ × _____ = _______ y = _________
i. A track in the woods is 300 yards long. Another track is 1,200 yards long. How many times longer is the second track than the first? _____ × _____ = _______ y = _________	**j.** Margaret has made 40 cups of jelly and she puts it in pint jars. How many jars will be filled? _____ × _____ = _______ y = _________

Multiplication Tables Review

Why is it important to learn your multiplication tables? Why couldn't you just use addition or other ways to find what is 6×9, 7×8, or 4×7? There are several reasons:

1. The knowledge of multiplication tables is needed for the opposite operation: division. Once you know the tables, you can do divisions such as $54 \div 6$ or $56 \div 7$ quickly in your head.
2. You also need to know the multiplication tables in order to perform long division.
3. Knowing the tables helps you be able to quickly simplify fractions. For example, you need to be able *to immediately* notice that both numbers in the fraction $\frac{56}{64}$ are in the table of 8.
4. Fraction addition and subtraction are very difficult if you don't know your tables by heart.
5. The tables are also necessary to be able to find factors and prime factorization of numbers.

1. Fill in the multiplication tables below and answer the questions.

$1 \times 5 =$	$7 \times 5 =$	$1 \times 10 =$	$7 \times 10 =$	$1 \times 11 =$	$7 \times 11 =$
$2 \times 5 =$	$8 \times 5 =$	$2 \times 10 =$	$8 \times 10 =$	$2 \times 11 =$	$8 \times 11 =$
$3 \times 5 =$	$9 \times 5 =$	$3 \times 10 =$	$9 \times 10 =$	$3 \times 11 =$	$9 \times 11 =$
$4 \times 5 =$	$10 \times 5 =$	$4 \times 10 =$	$10 \times 10 =$	$4 \times 11 =$	$10 \times 11 =$
$5 \times 5 =$	$11 \times 5 =$	$5 \times 10 =$	$11 \times 10 =$	$5 \times 11 =$	$11 \times 11 =$
$6 \times 5 =$	$12 \times 5 =$	$6 \times 10 =$	$12 \times 10 =$	$6 \times 11 =$	$12 \times 11 =$

To find a number times 5, first multiply that number by 10, and take half of that. So for 7×5, first go $7 \times 10 = 70$ and take half of that.

Elevens are as easy as pie!

2. What same answers do you find in the tables of 5 and 10? Why?

$1 \times 2 =$	$7 \times 2 =$	$1 \times 4 =$	$7 \times 4 =$	$1 \times 8 =$	$7 \times 8 =$
$2 \times 2 =$	$8 \times 2 =$	$2 \times 4 =$	$8 \times 4 =$	$2 \times 8 =$	$8 \times 8 =$
$3 \times 2 =$	$9 \times 2 =$	$3 \times 4 =$	$9 \times 4 =$	$3 \times 8 =$	$9 \times 8 =$
$4 \times 2 =$	$10 \times 2 =$	$4 \times 4 =$	$10 \times 4 =$	$4 \times 8 =$	$10 \times 8 =$
$5 \times 2 =$	$11 \times 2 =$	$5 \times 4 =$	$11 \times 4 =$	$5 \times 8 =$	$11 \times 8 =$
$6 \times 2 =$	$12 \times 2 =$	$6 \times 4 =$	$12 \times 4 =$	$6 \times 8 =$	$12 \times 8 =$

What same answers (products) do you find in the tables of 2, 4, and 8?

To find a number times 4, you can double it twice:

Example. $7 \times 4 = ??$
Double 7 is 14, then double that to get 28.

You can double a number three times to find these. For example, to find 6×8, find double 6, and double that, then double that.

5, 6, 7, 8 — fifty-six is 7 times 8.

Color ones digits one color and tens digits another. You will see a pattern.

3. Fill in the multiplication tables below and answer the questions.

$1 \times 3 =$	$7 \times 3 =$	$1 \times 6 =$	$7 \times 6 =$	$1 \times 9 =$	$7 \times 9 =$
$2 \times 3 =$	$8 \times 3 =$	$2 \times 6 =$	$8 \times 6 =$	$2 \times 9 =$	$8 \times 9 =$
$3 \times 3 =$	$9 \times 3 =$	$3 \times 6 =$	$9 \times 6 =$	$3 \times 9 =$	$9 \times 9 =$
$4 \times 3 =$	$10 \times 3 =$	$4 \times 6 =$	$10 \times 6 =$	$4 \times 9 =$	$10 \times 9 =$
$5 \times 3 =$	$11 \times 3 =$	$5 \times 6 =$	$11 \times 6 =$	$5 \times 9 =$	$11 \times 9 =$
$6 \times 3 =$	$12 \times 3 =$	$6 \times 6 =$	$12 \times 6 =$	$6 \times 9 =$	$12 \times 9 =$

To find a number times 6, you can double the corresponding one from the table of 3:

$6 \times 7 = ??$ Find 3×7, and double that.

What same products do you find in the tables of 3 and 6?

Why is that?

The table of 9 has **special things**!

Color all the ones digits yellow (of the answers). Color all the tens digits red (of the answers).

Add the digits of each answer. What do you notice?

4. These are harder ones... but remember, you can change the order of multiplication. 8×7 is the same as 7×8, which is 56.

$1 \times 7 =$	$7 \times 7 =$	$1 \times 12 =$	$7 \times 12 =$
$2 \times 7 =$	$8 \times 7 =$	$2 \times 12 =$	$8 \times 12 =$
$3 \times 7 =$	$9 \times 7 =$	$3 \times 12 =$	$9 \times 12 =$
$4 \times 7 =$	$10 \times 7 =$	$4 \times 12 =$	$10 \times 12 =$
$5 \times 7 =$	$11 \times 7 =$	$5 \times 12 =$	$11 \times 12 =$
$6 \times 7 =$	$12 \times 7 =$	$6 \times 12 =$	$12 \times 12 =$

5. It is time to test your knowledge with missing factor problems!

a. ____ $\times 7 = 49$ ____ $\times 7 = 28$ ____ $\times 7 = 56$	**b.** ____ $\times 6 = 48$ ____ $\times 6 = 30$ ____ $\times 6 = 54$	**c.** ____ $\times 8 = 64$ ____ $\times 8 = 48$ ____ $\times 8 = 56$	**d.** ____ $\times 9 = 72$ ____ $\times 9 = 54$ ____ $\times 9 = 63$
e. ____ $\times 5 = 45$ ____ $\times 5 = 35$ ____ $\times 5 = 40$	**f.** ____ $\times 3 = 27$ ____ $\times 3 = 18$ ____ $\times 3 = 21$	**g.** ____ $\times 4 = 28$ ____ $\times 4 = 36$ ____ $\times 4 = 32$	**h.** ____ $\times 2 = 18$ ____ $\times 2 = 16$ ____ $\times 2 = 12$
i. ____ $\times 7 = 35$ ____ $\times 7 = 63$ ____ $\times 7 = 21$	**j.** ____ $\times 5 = 60$ ____ $\times 5 = 25$ ____ $\times 5 = 30$	**k.** ____ $\times 6 = 36$ ____ $\times 6 = 72$ ____ $\times 6 = 42$	**l.** ____ $\times 8 = 72$ ____ $\times 8 = 16$ ____ $\times 8 = 32$

6. Fill in the table.

×	0	1	2	3	4	5	6	7	8	9	10	11	12
0													
1													
2													
3													
4													
5													
6													
7													
8													
9													
10													
11													
12													

7. A school hired a bus and eight minivans to take the 90 students to a swimming pool.
All the minivans were full, with 7 students in each.
How many students went in the bus?

8. Let's practice the order of operations again.

a. $4 \times 7 + 5 =$ ______	**c.** $4 \times (7 - 5) =$ ______	**e.** $(4 + 5) \times (5 + 2) =$ ______
b. $2 \times (5 + 6) + 4 =$ ______	**d.** $100 - 5 \times 6 =$ ______	**f.** $70 - (5 + 6) \times 4 =$ ______

9. Fill in the missing numbers so that both sides of the equals sign "=" have the same value.

Example: $2 \times 12 = 8 \times$ _3_ because $24 = 24$	**a.** $2 \times 6 = 4 \times$ ______ $12 = 12$	**b.** $6 \times 6 = 4 \times$ ______
c. $3 \times 10 = 6 \times$ ______	**d.** $2 \times 20 = 10 \times$ ______	**e.** $5 \times 12 = 6 \times$ ______

Scales Puzzles

This is a pan balance or scales. You put things into the two "pans," and the heavier pan will go down, like in a seesaw. If the two things weigh the same, the balance stays balanced.

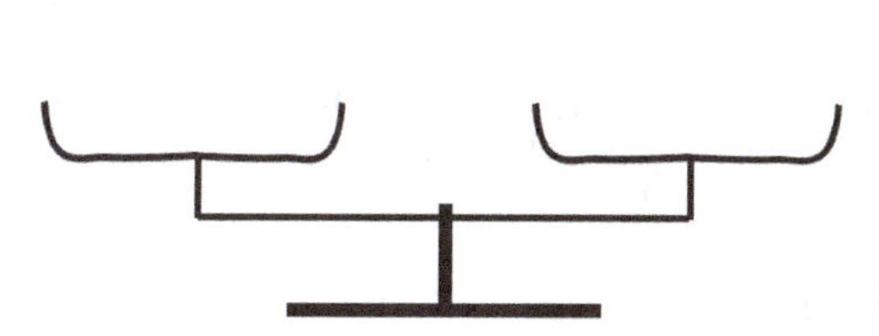

We can use the pan balance to model simple equations with an unknown. In this lesson you will solve many equations with its help.

1. Solve how much each geometric shape "weighs." You can imagine the weights being so many pounds or kilograms, if it helps.

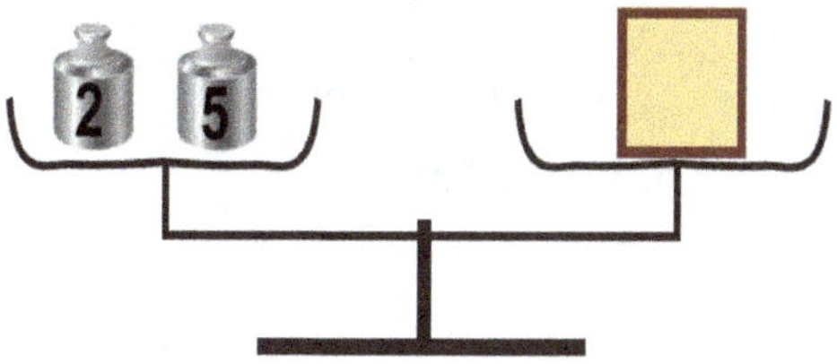

a. The square weighs _______	**b.** The square weighs _______
c. One ball weighs _______	**d.** One rectangle weighs _______
e. One pentagon weighs _______	**f.** One oval weighs _______
g. One square weighs _______	**h.** One square weighs _______

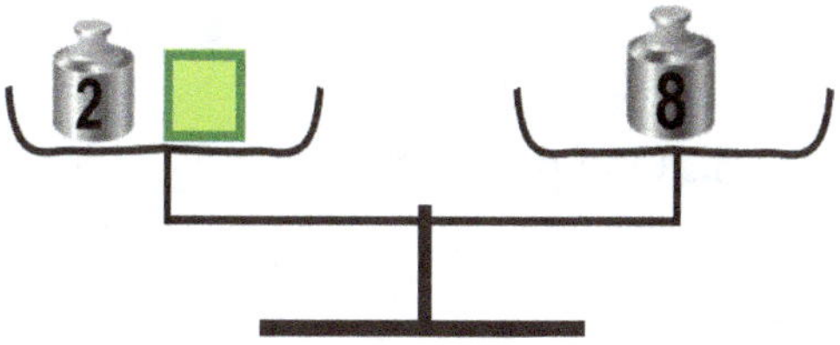

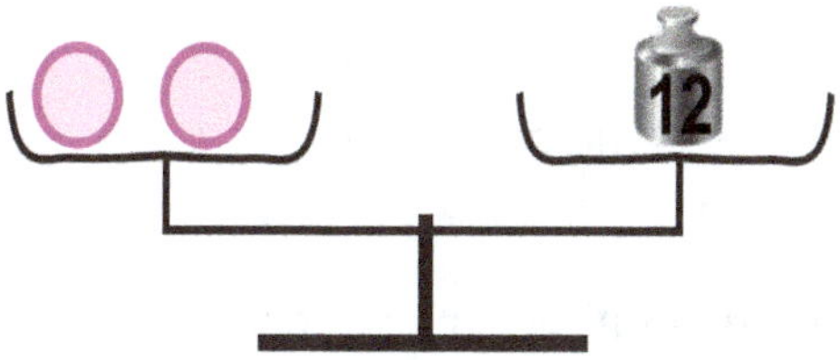

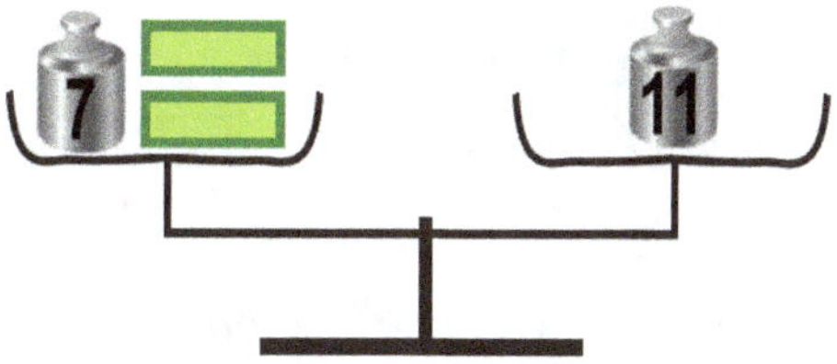

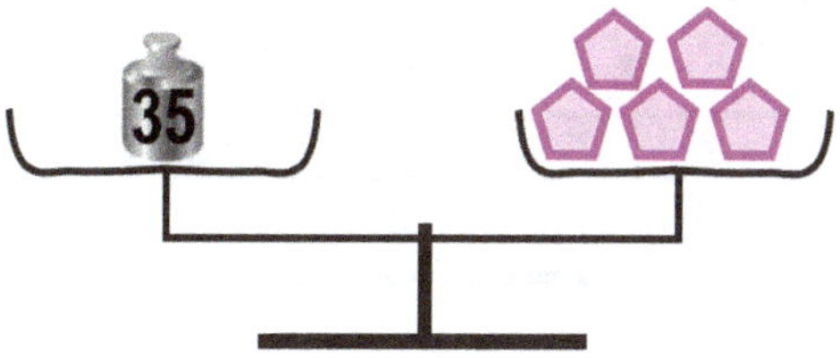

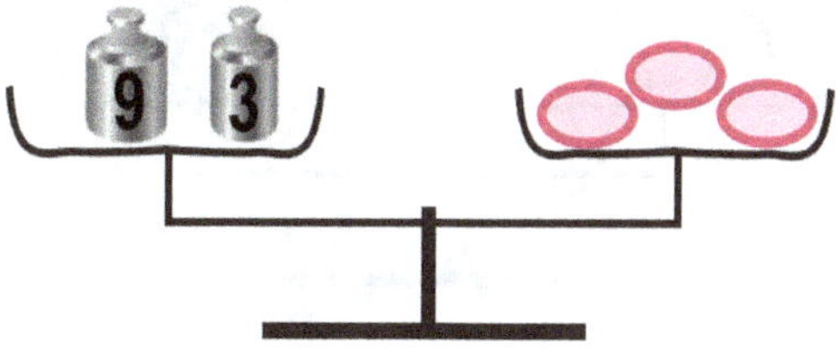

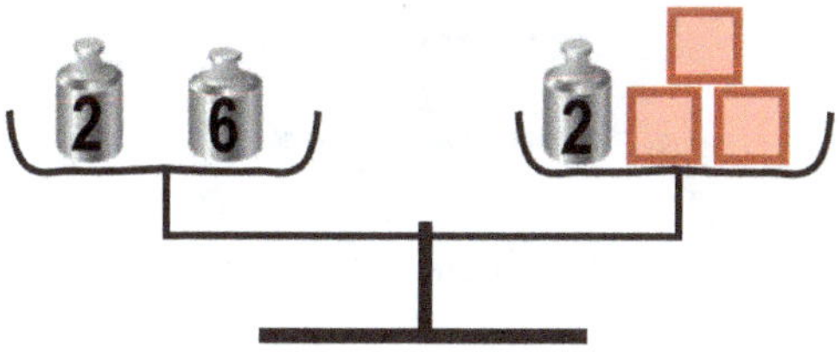

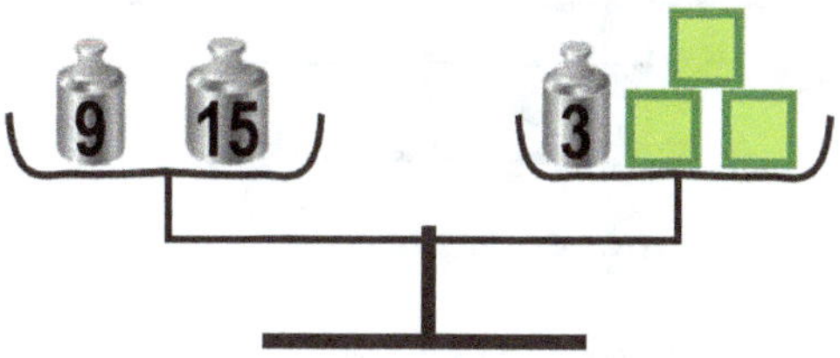

If there are "unknown shapes" on both sides, use this "trick":

Cross out the same amount of unknown shapes from both sides.

That way the balance will continue to stay balanced!

We cross out two diamonds from both sides.
Then we see that *three* diamonds weigh 15.
This of course means that one diamond weighs 5.

2. Solve the pan balance equations.

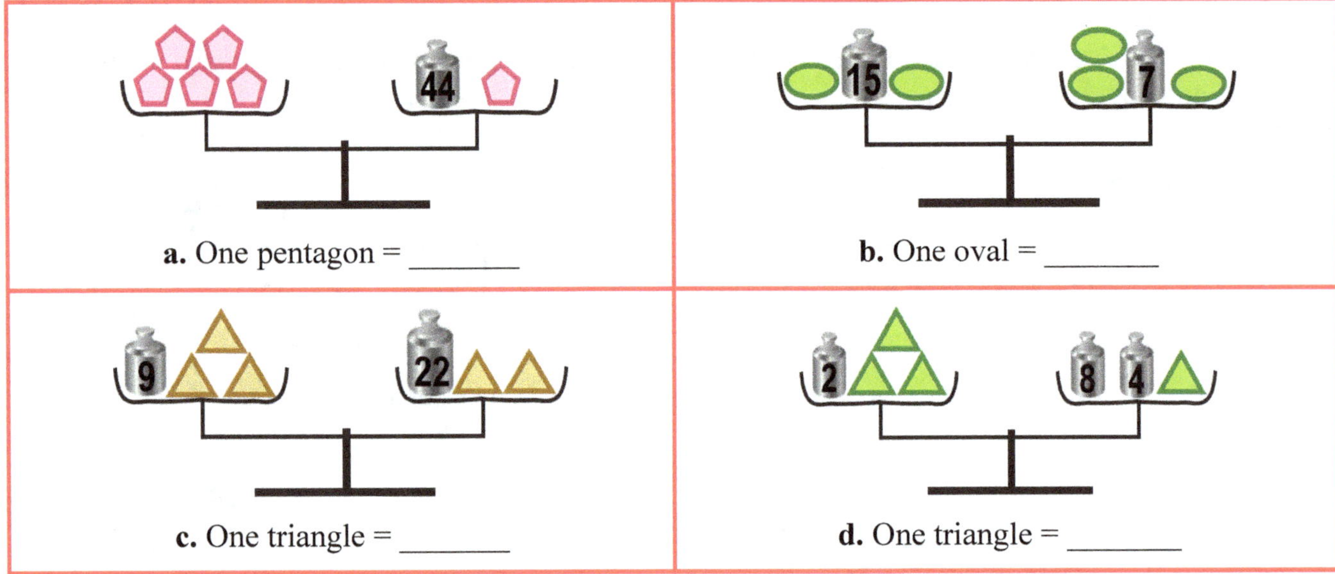

a. One pentagon = _______

b. One oval = _______

c. One triangle = _______

d. One triangle = _______

3. Solve. These are trickier. Use *both* balances to figure out the *two* unknown shapes. Guess and check! See the answer key for a hint.

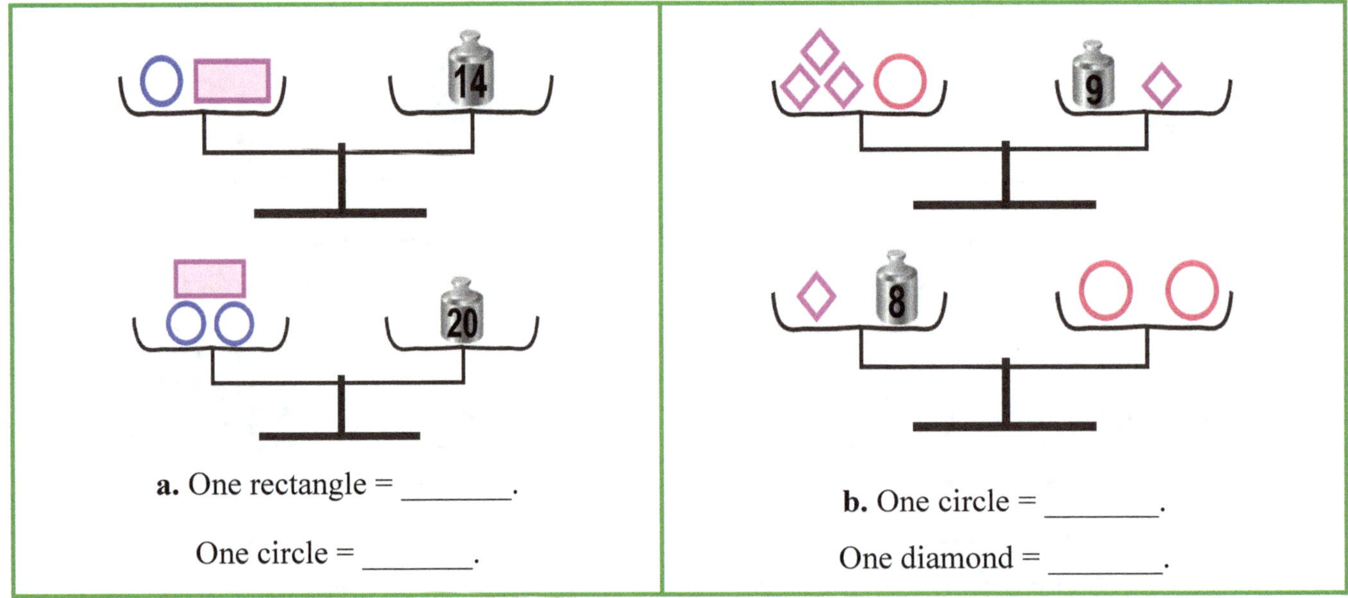

a. One rectangle = _______.

One circle = _______.

b. One circle = _______.

One diamond = _______.

4. These are double scales. Each scale on one side is like a puzzle in itself—so solve that first!

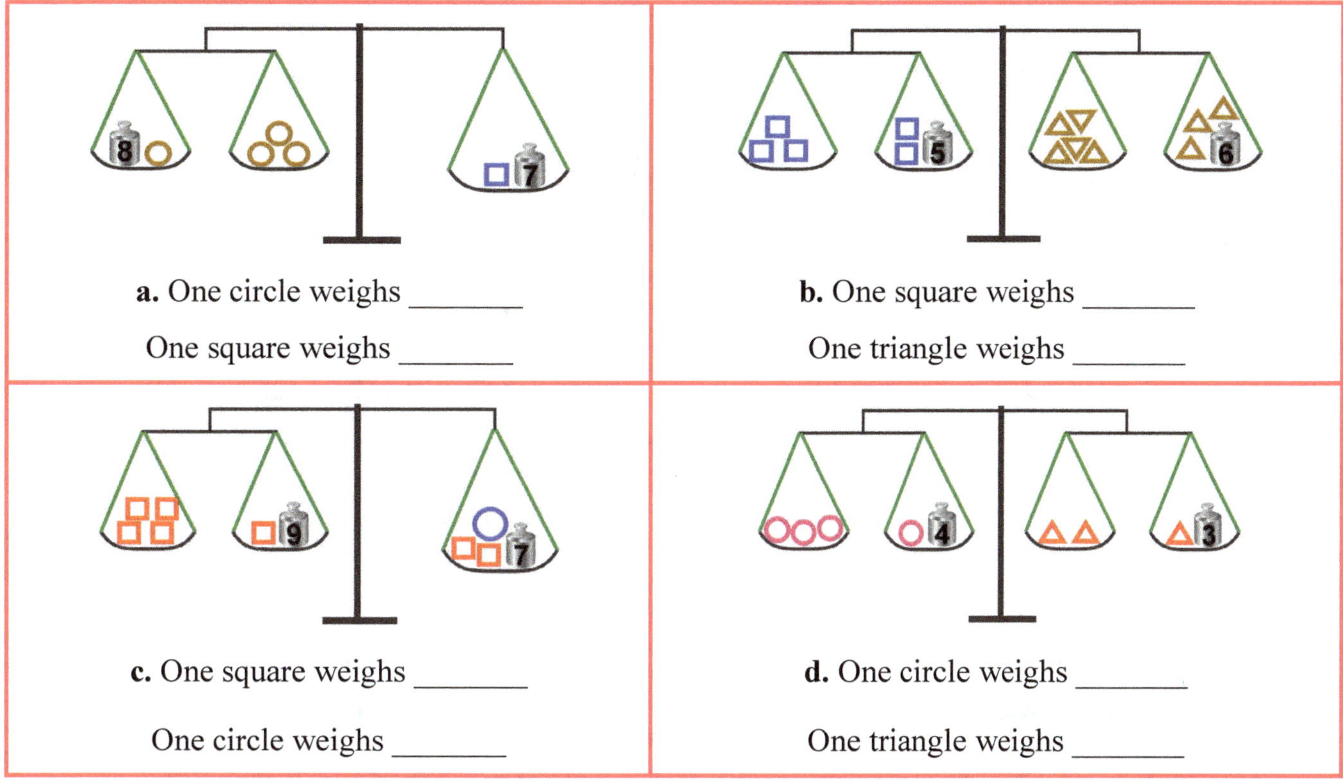

In mathematics, the equal sign "=" is like a scales that is balanced. Something is on the right side, something else is on the left side, and the sides are equal or "balanced."

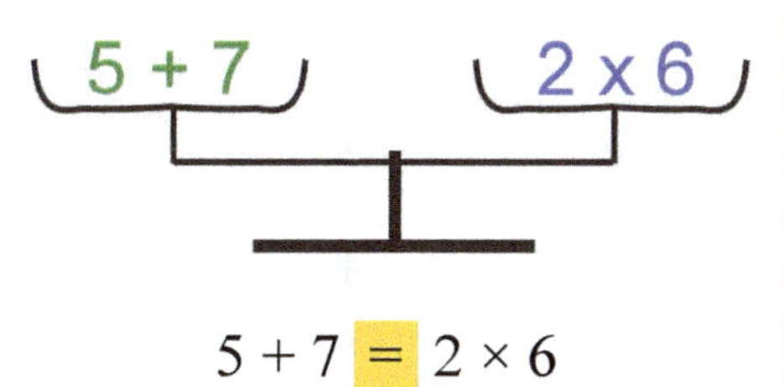

$5 + 7 = 2 \times 6$

5. Find the numbers that go on the empty lines. These equations are like little puzzles!

a. 78 + ____ = 148	**b.** 7 + 6 + 6 = _____ − 10	**c.** 2 × 50 = 40 + _____
d. 160 = ___ + 9	**e.** 5 + 5 + 5 + ____ = 2 × 12	**f.** 7 × 6 = 2 × _____
g. 50 − ____ = 32	**h.** 16 + 19 = 2 × ____ + 1	**i.** 4 × 6 − 7 = 2 × ___ + 1

On the next page you will find pictures of empty scales. You can print out the page and make up your own problems but be careful! If you just make random problems, the solutions are likely to be fractions. See also:

http://mste.illinois.edu/users/pavel/java/balance/

https://www.transum.org/Software/SW/Starter_of_the_day/Students/Stable_Scales_Quiz.asp – weighing scales game that practices algebraic reasoning

https://www.nctm.org/Classroom-Resources/Illuminations/Interactives/Pan-Balance----Shapes/ – an interactive pan balance with shapes.

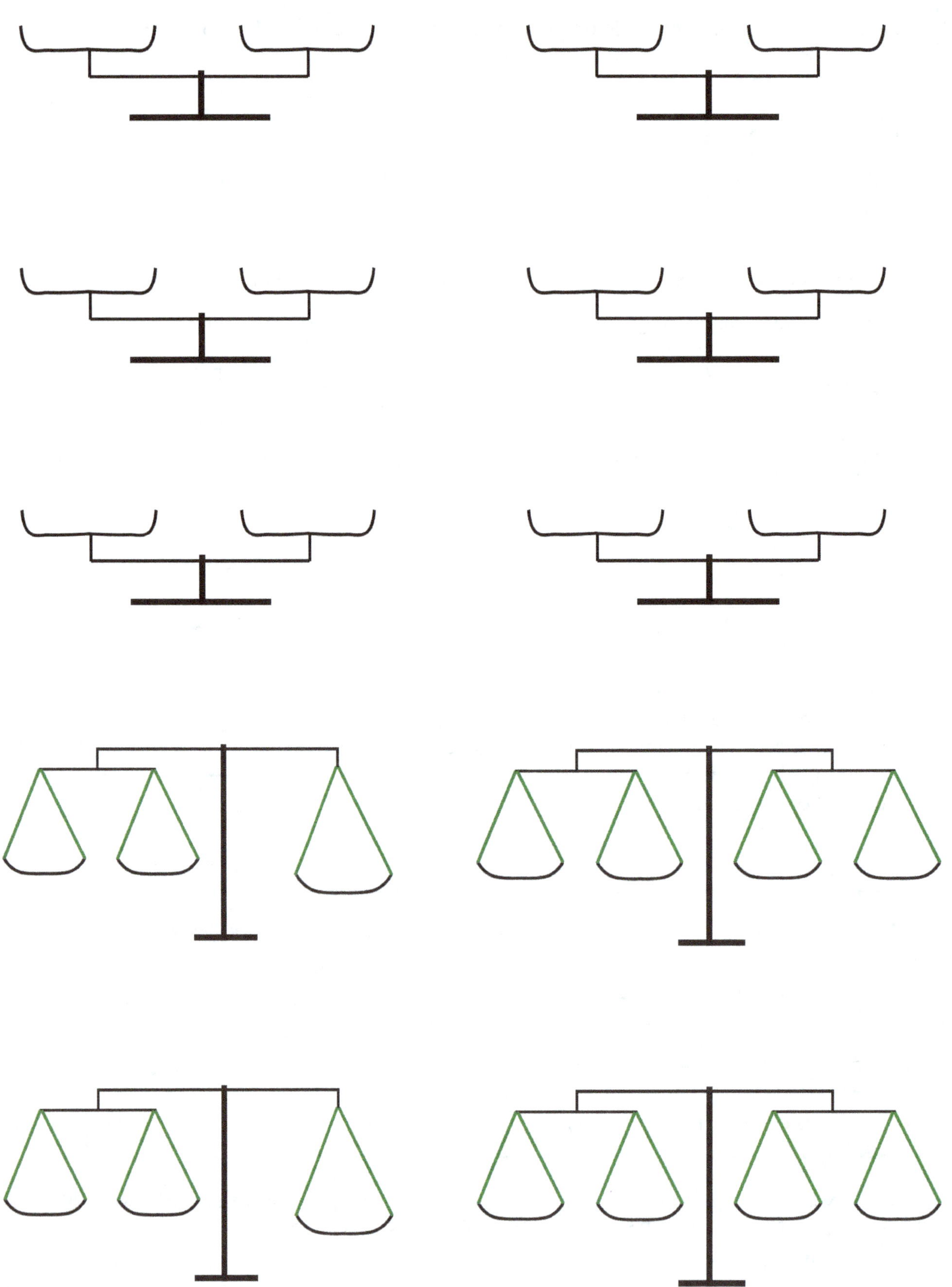

Multiplying by Whole Tens and Hundreds

We have studied the SHORTCUTS for multiplying any number by 10, 100, or 1,000:		
To multiply any number by **10**, just tag **ONE zero** to the end. To multiply any number by **100**, just tag **TWO zeros** to the end. To multiply any number by **1,000**, just tag **THREE zeros** to the end.		
1**0** × 481 = 4,81**0**	1**00** × 47 = 4,7**00**	1**000** × 578 = 578**,000**
Note especially what happens when the number you multiply already ends in a zero or zeros. The rule works the same way, and you *still* have to tag the zero or zeros.		
1**0** × 800 = 800**0**	1**00** × 6,600 = 660,0**00**	1**000** × 40 = 40**,000**

1. Multiply.

a. 10 × 315 = _______	**b.** 100 × 6,200 = _______	**c.** 1,000 × 250 = _______
3,560 × 10 = _______	10 × 1,200 = _______	38 × 1,000 = _______
35 × 100 = _______	100 × 130 = _______	10 × 5,000 = _______

Shortcut for multiplying by 20 or 200 (You can probably guess this one!)	
What is 20 × 14? First solve the problem without the zero in 20: 2 × 14 = 28. Next, tag a zero to the answer, 28, and you get 280. So, 20 × 14 = 280.	**What is 200 × 31?** First solve the problem without the zeros: 2 × 31 = 62. Next, just *two* zeros to the result, 62, to get 6,200. In other words, 200 × 31 = 6,200.

2. Now try it! Multiply by 20 and 200.

a.	**b.**	**c.**	**d.**
20 × 8 = ______	200 × 7 = _______	20 × 12 = _______	20 × 16 = _________
4 × 20 = ______	5 × 200 = _______	35 × 20 = _______	42 × 200 = _________
20 × 5 = ______	11 × 200 = _______	200 × 9 = _______	54 × 20 = _________

Why does the shortcut work? It is based on the fact that we can multiply numbers in any order.	
When multiplying any number by 20, we can write the 20 as 10 × 2. For example: $\underline{20} \times 14 = \underline{10 \times 2} \times 14$ In that problem, first multiply 2 × 14 = 28. Then the problem becomes 10 × 28, which equals 280. Notice again how we did it: $\underline{20} \times 14$ $= \underline{10 \times 2} \times 14$ $= 10 \times 28$ $= 280$	Let's try the same idea with 200. We will write 200 as 100 × 2. For example: $\underline{200} \times 31 = \underline{100 \times 2} \times 31$ In that problem, first multiply 2 × 31 = 62. The problem now becomes 100 × 62, which is 6,200. Notice again how it was done: $\underline{200} \times 31$ $= \underline{100 \times 2} \times 31$ $= 100 \times 62$ $= 6{,}200$

3. Try it yourself! Fill in.

a. 20 × 7	**b.** 20 × 5	**c.** 200 × 8	**d.** 200 × 25
= ______ × 2 × 7	= ______ × 2 × 5	= ______ × 2 × 8	= ______ × 2 × 25
= 10 × ______	= 10 × ______	= 100 × ______	= 100 × ______
= _________	= _________	= _________	= _________

4. Mark's shed measures 20 ft by 15 ft. Write and solve a number sentence for its area. ("A" means area.)

A = ________________________________

Hint: To calculate the area of a rectangle, multiply its two sides.

5. Write a number sentence to find the area of Mark's driveway, and solve it.

200 ft

15 ft

A = ________________________________

6. Mark was told he needed four truckloads of gravel to cover his driveway. *One* truckload costs 5 × $20 plus $30 for the delivery. How much will it cost him to cover the driveway with gravel?

SHORTCUT for multiplying by whole tens and whole hundreds			
The same principle works if you multiply by whole tens (30, 40, 50, 60, 70, 80, or 90): simply multiply by 3, 4, 5, 6, 7, 8, or 9, and then tag a zero to the result. Similarly, if you multiply by some whole hundred, first solve the multiplication without the two zeros of the hundreds, and then tag two zeros to the result.			
50 × 8 = 400	90 × 11 = 990	300 × 8 = 2,400	12 × 800 = 9,600

7. Multiply.

a. 40 × 3 = ________ 8 × 20 = ________	**b.** 70 × 6 = ________ 50 × 11 = ________	**c.** 80 × 9 = ________ 30 × 15 = ________
d. 60 × 11 = ________ 12 × 40 = ________	**e.** 200 × 9 = ________ 7 × 400 = ________	**f.** 700 × 6 = ________ 600 × 11 = ________
g. 200 × 12 = ________ 15 × 300 = ________	**h.** 3 × 1100 = ________ 8 × 900 = ________	**i.** 11 × 120 = ________ 8 × 300 = ________

It even works this way:		
To multiply 40 × 70, simply multiply 4 × 7, and tag two zeros to the result: 40 × 70 = 2,800	To multiply 600 × 40, simply multiply 6 × 4, and tag three zeros to the result: 600 × 40 = 24,000	To multiply 700 × 800, simply multiply 7 × 8, and tag four zeros to the result. 700 × 800 = 560,000

8. Multiply.

a. 20 × 90 = ________ 70 × 300 = ________	**b.** 60 × 80 = ________ 30 × 900 = ________	**c.** 400 × 50 = ________ 200 × 200 = ________
d. 80 × 800 = ________ 200 × 500 = ________	**e.** 100 × 100 = ________ 40 × 30 = ________	**f.** 800 × 300 = ________ 90 × 1100 = ________

Write a number sentence for each question.

9. One hour has ______ minutes.

How many minutes are in 12 hours? ______________________________

How many minutes are in 24 hours? ______________________________

10. One hour has ______ minutes, and one minute has ______ seconds.

How many seconds are there in one hour? ______________________________

11. Ed earns $30 per hour.

a. How much will he earn in an 8-hour workday? ______________________________

b. How much will he earn in a 40-hour workweek? ______________________________

c. How many days will he need to work in order to earn more than $1,000?

____

12. Find the missing factor. Think "backwards": how many zeros do you need?

a. ________ × 3 = 360 ________ × 50 = 450	**b.** 40 × ________ = 320 5 × ________ = 600	**c.** ________ × 40 = 400 ________ × 2 = 180
d. ________ × 30 = 4,800 ________ × 200 = 1,800	**e.** 40 × ________ = 2,000 6 × ________ = 4,200	**f.** ________ × 800 = 56,000 ________ × 20 = 12,000

Puzzle Corner

John wanted to prove that 40 × 70 is indeed 2,800 by breaking the multiplication into smaller parts. He wrote 40 as 4 × 10 and 70 as 7 × 10, and then multiplied in a different order:

$$40 \times 70 = 4 \times 10 \times 7 \times 10$$

$$= 10 \times 10 \times (4 \times 7) = 100 \times 28 = 2{,}800.$$

Do the same, and prove that 600 × 50 is indeed 30,000.

Multiply in Parts 1

Example 1. To multiply 3 × 46, break 46 into two parts: 40 and 6.

Then multiply those two parts separately by 3:
3 × 40 is 120, and 3 × 6 is 18.

Lastly add these two partial results: 120 + 18 = 138.

Example 2. This illustration shows the same thing, 3 × 46, using bundles of ten.

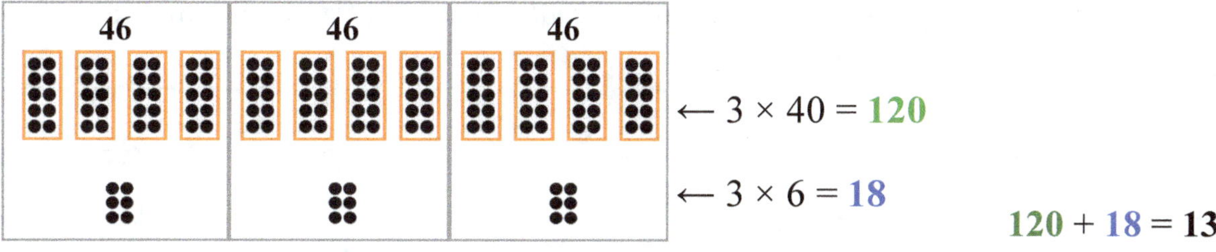

Study these examples. Multiply the tens and ones separately, then add:

8 × 13	**5 × 24**	**7 × 68**
(10 + 3)	(20 + 4)	(60 + 8)
8 × 10 and 8 × 3	5 × 20 and 5 × 4	7 × 60 and 7 × 8
80 and 24	100 and 20	420 and 56
= 104	= 120	= 476

1. Multiply the tens and ones separately. Then add to get the final answer.

a. 6 × 27	**b. 5 × 83**	**c. 9 × 34**
(20 + 7)	(+)	(+)
6 × ______ and 6 × _____	5 × ______ and 5 × _____	9 × ______ and 9 × _____
______ and ______	______ and ______	______ and ______
= __________	= __________	= __________
d. 3 × 99	**e. 7 × 65**	**f. 4 × 58**
3 × ______ and 3 × _____	7 × ______ and 7 × _____	4 × ______ and 4 × _____
______ and ______	______ and ______	______ and ______
= __________	= __________	= __________

Example 3. The picture shows the area of a rectangle with sides 8 and 24. It is also divided into two rectangles.

The area of the WHOLE rectangle is 8 × 24 square units. We can find 8 × 24 by calculating the areas of the *two* rectangles, and adding.

The area of the first rectangle is 8 × 20 = 160 square units. The area of the second rectangle is 8 × 4 = 32 square units.

Then, the area of the whole rectangle is the sum 160 + 32 = 192 square units.

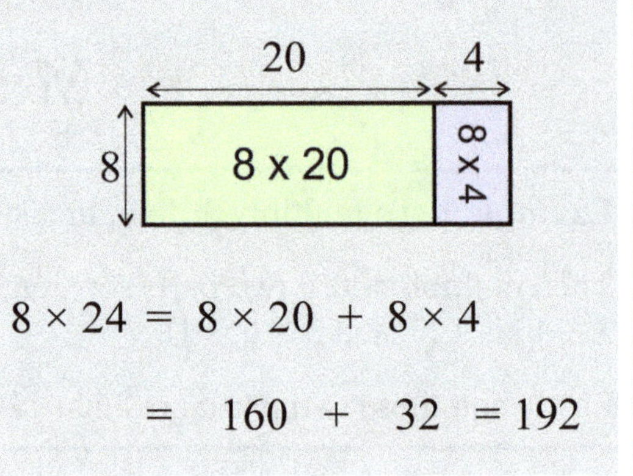

2. Fill in the missing numbers. Write the area of the *whole* rectangle as a SUM of the areas of the *smaller* rectangles. Also find the total area.

a. ___ × ______ = ___ × ______ + ___ × ___

= ______ + ______ = ________

10 7

9

b. ___ × ______ = ___ × ______ + ___ × ___

= ______ + ______ = ________

20 9

6

c. ___ × ______ = ___ × ______ + ___ × ___

= ______ + ______ = ________

30 3

7

3. It is your turn to draw. Draw a two-part rectangle to illustrate the multiplications, like in the previous problem. You don't have to draw accurately; a sketch is good enough.

a. 7 × 16 = ___ × ______ + ___ × ___

=

b. 5 × 21 = ___ × ______ + ___ × ___

=

c. 8 × 34 = ___ × ______ + ___ × ___

=

4. Break the second factor into tens and ones. Multiply separately, and add.

a. 6 × 19	**b. 3 × 73**	**c. 4 × 67**
6 × 10 → 60 6 × 9 → + 54 114	3 × ____ → 3 × ____ → + ____	
d. 5 × 92	**e. 9 × 33**	**f. 7 × 47**

5. Multiply in parts. You can write the partial products under the problems, if you wish.

a. 5 × 13 = ______	**b.** 9 × 15 = ______	**c.** 5 × 33 = ______
d. 8 × 21 = ______	**e.** 4 × 22 = ______	**f.** 7 × 51 = ______

6. Compare. Write < , > , or = in the boxes.

a. 10 × 10 ☐ 9 × 11　　**b.** 6 × 12 ☐ 5 × 14　　**c.** 8 × 22 ☐ 5 × 27

7. Solve. Write a number sentence for each problem, *not* just the answer.

a. Jack bought eight shirts for $14 each. What was the total cost?

____

b. Mary and Harry set up nine rows of seats in the school auditorium, with 14 seats in each row. After that, they still had 56 seats left in the storage that they didn't use. How many seats are there in total?

____

c. A small hammer costs $17. Another, much better one, costs three times as much. Find the cost of the more expensive hammer.

____

Multiply in Parts 2

Multiplying in parts (partial products) works with larger numbers, too:

7×329	$5 \times 2{,}395$
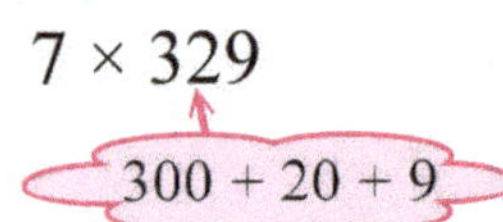	2,000 + 300 + 90 + 5
(7×300) and (7×20) and (7×9)	$(5 \times 2{,}000) + (5 \times 300) + (5 \times 90) + (5 \times 5)$
7×300 is 2,100. 7×20 is 140. $7 \times 9 = 63$.	$5 \times 2{,}000$ is 10,000. 5×300 is 1,500. 5×90 is 450. And 5×5 is 25.
Lastly, add the partial results: 2100 + 140 + 63 = 2,303	Lastly, add the partial results: 10000 + 1500 + 450 + 25 = 11,975

1. Multiply in parts. Then add to get the final answer. Use the grids for additions.

a. 3 × 127
(100 + 20 + 7)

3 × ______ and 3 × ______ and 3 × ____

______ and ______ and ____

= ___________

b. 5 × 243
()

5 × ______ and 5 × ______ and 5 × ____

______ and ______ and ____

= ___________

c. 4 × 6,507
(   )

4 × ______ + 4 × ______ + 4 × ____

________ + ________ + ______

= ___________

d. 5 × 4,813

5 × ______ + 5 × ______ + 5 × 10 + 5 × ___

________ + ________ + ______ + ______

= ___________

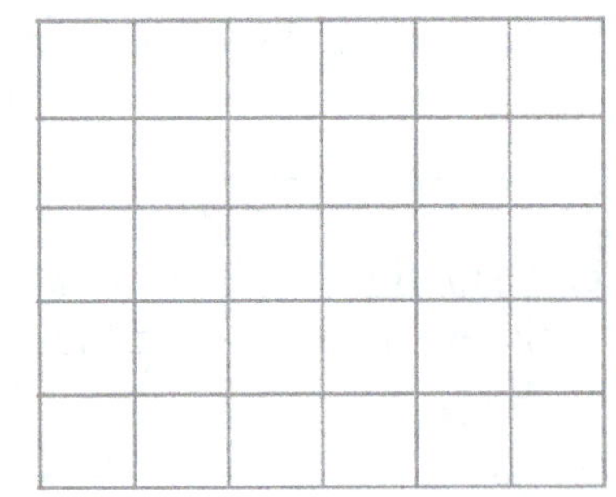

2. Break the second factor into thousands, hundreds, tens, and ones. Multiply separately, and add.

a. 4 × 128	**b.** 8 × 151	**c.** 3 × 452
4 × 100 → 4 0 0 4 × 20 → 8 0 4 × 8 → + 3 2	+	+
d. 6 × 3,217	**e.** 8 × 2,552	**f.** 6 × 1,098
+	+	+

3. Solve. Write a number sentence for each problem, *not* just the answer.

a. Dad pays $138 in rent each month. How much does he pay in rent in half a year?

b. Find the perimeter of a square with 255 cm sides.

c. One roll contains 56 cm of material, and another roll contains five times as much. Find how much material both rolls contain in total.

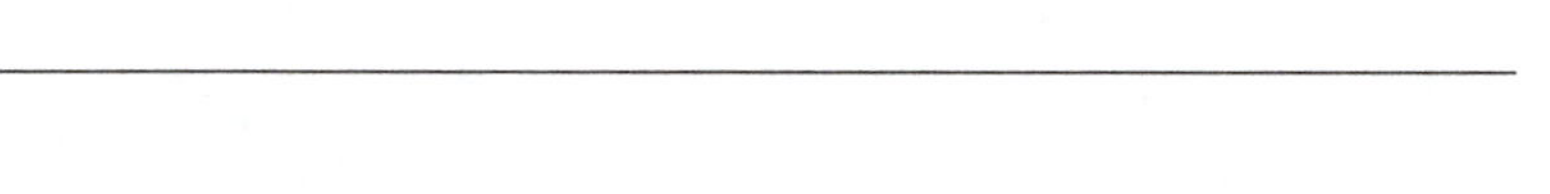

Multiply in Parts – Area Model

1. Write the area of the *whole* rectangle as a SUM of the areas of the *smaller* rectangles. Also find the total area. Use the grid for the final addition.

a.

8 × 127

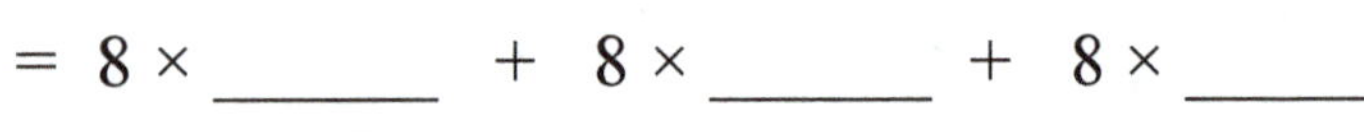

= 8 × ______ + 8 × ______ + 8 × ______

The total area is ____________ square units.

b.

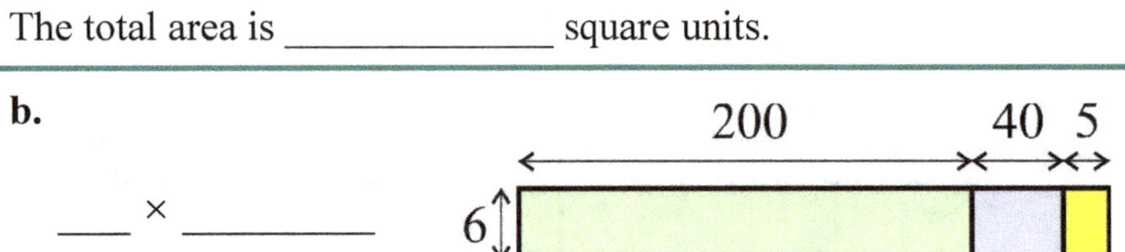

___ × ________

= ___ × ______ + ___ × ______ + ___ × ______

The total area is ____________ square units.

c.

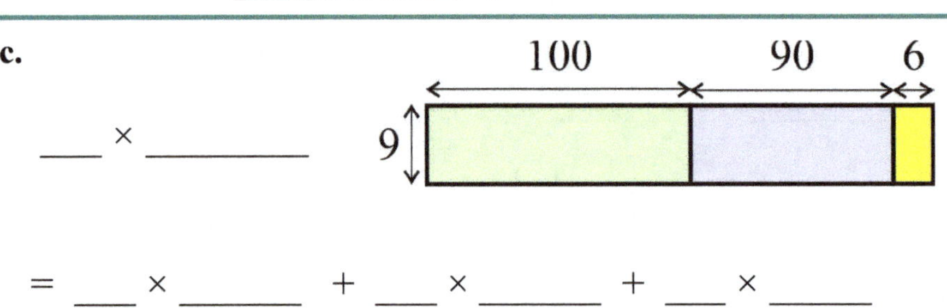

___ × ________

= ___ × ______ + ___ × ______ + ___ × ______

The total area is ____________ square units.

Remember? The expressions (number sentences) on the left and right sides of the "=" sign have *an equal value*:	2 × 6 = 3 × 4 12 = 12	18 − 3 = 5 × 3 15 = 15

2. Fill in the missing numbers to make the expressions equal.

a. 6 × 6 = 9 × _______

b. _____ × 10 = 5 × 24

c. 20 + _____ = 4 × 10

d. 6000 = 30 × _________

e. 120 − 75 = 5 × ______

f. ______ + 750 = 5 × 300

3. It is your turn to draw. Draw a *three*-part rectangle to illustrate the multiplications, as in the previous problems. You don't have to draw accurately; a sketch is good enough. Find the area of each part and the total area. Use the grids for the additions.

a. 7×153

Areas of the parts:

Total area:

b. 5×218

Areas of the parts:

Total area:

c. 8×376

Areas of the parts:

Total area:

4. Solve. Write number sentence(s) on the empty lines to show your work.

a. Susie orders roses for her flower shop in bunches of six dozen (72 flowers) at a time. She places an order once a week. How many roses will Susie order in total for five weeks?

____

____

b. One bunch of six dozen roses costs her $70. What is the total cost for five weeks of orders for roses?

____

____

Multiplying Money Amounts

We can also multiply in parts with money amounts.	
3 × $1.70 Think of $1.70 as $1 and 70¢. Multiply separately: 3 × $1 is $3.00. 3 × 70¢ is 210¢ or $2.10. Lastly add: $3.00 + $2.10 = $5.10.	**8 × $4.28** Multiply in parts: 8 × $4 is $32. Next, 8 × 20¢ is 160¢ or $1.60. Then, 8 × 8¢ is 64¢. Lastly add (on the right). $32.00 $0.64 + $1.60 $34.24

1. Multiply in parts, and lastly add.

a. 6 × $11.85	**b. 5 × $2.93**
6 × $11 → 6 × $0.80 → 6 × $0.05 → + ______	+ ______
c. 7 × $3.75	**d. 8 × $10.95**
+ ______	+ ______

2. Break the money amounts into dollars and cents. Multiply separately, and lastly add.

a. 6 × $2.80	**b. 5 × $4.70**
______ + ______ = (6 × $2) (6 × $0.80)	______ + ______ = (5 × $4) (5 × $0.70)
c. 4 × $12.50	**d. 7 × $5.61**

3. **a.** Sandra bought five books for $2.70 each.
What was the total cost?

b. She paid with $20. What was her change?

4. **a.** A ticket to a concert costs $23.50. What is the total cost for tickets for a family of four?

b. What is their change from $100?

5. Continue the patterns according to the instructions.

a. Start at 80. Add 40 each time:	**b.** Start at 42,000. Subtract 3,000 each time:	**c.** Start at 1. Add 5 each time:
1 2 0	______	______
______	______	______
______	______	______
______	______	______
______	______	______
______	______	______
______	______	______
What does this pattern remind you of?	What does this pattern remind you of?	______
		______

6. Notice the pattern of odd and even numbers in part (c). Can you explain why this happens?

Estimating in Multiplication

If you don't need an exact result, you can estimate. To estimate the result of a multiplication, round some or all of the factors so that you can easily multiply *mentally*.

Estimate 8 × 189.	**Estimate 42 × 78.**	**Estimate 21 × $4.52.**
189 can be rounded to 200. The estimated product is 8 × 200 = 1,600.	42 ≈ 40 and 78 ≈ 80. The estimated product is 40 × 80 = 3,200.	First round the numbers to 20 and $4.50. Then, since 2 × $4.50 is nine dollars, 20 × $4.50 is ten times that, or $90.

1. Estimate the results by rounding one or both factors. Don't round both numbers if you can multiply in your head just by rounding one factor.

a. 5 × 69 ≈ ____ × ______ = ________	**b.** 11 × 58 ≈ _____ × ______ = _______	**c.** 119 × 8 ≈ _______ × ____ = ________
d. 27 × 52 ≈ ______ × ______ = _______	**e.** 7 × $4.15 ≈ ____ × _____ = ________	**f.** 8 × $11.79 ≈ ____ × ______ = _______
g. 25 × $42.50 ≈ _____ × _______ = _______	**h.** 9 × 17 ≈ ____ × _____ = ________	**i.** 63 × 897 ≈ _____ × _______ = _______

2. Estimate the total cost. Round one or both numbers so that you can multiply in your head. Write a number sentence to show your multiplication with rounded numbers.

a. 24 chairs at $44.95 per chair	**b.** 512 Popsicles at 19¢ each
c. 210 meters of wire at $1.49 per meter	**d.** Six tennis balls that cost $3.37 each and two rackets that cost $11.90 each.

Example. If each bus can seat 57 passengers, how many buses do you need to seat 450 people?

Let's think:

One bus seats 57 passengers. Two buses seat 114 passengers.	Ten buses seat 570 passengers. Eight buses seat 8×57 passengers.	*Etc.*

So how many buses will we need for our answer to be 450 or a little more?

This problem could be solved by division ($450 \div 57$), but it is easier to **use estimation and multiplication**. Round 57 to 60, and quickly calculate in your head this way:

$7 \times 60 = 420$ and $8 \times 60 = 480$. It *looks like* 8 buses are needed for 450 people.

Lastly, let's check our answer using the exact number 57:
$8 \times 57 = 400 + 56 = 456$, so yes, eight buses are enough to transport 450 people.

3. Solve the problems using estimation.

a. An advertisement in a newspaper costs $349.
How many ads can Bill buy with $2,000?

b. It costs $2.85 per hour to skate at a skating rink. Sandra has $25. How many whole hours can she afford to skate?

c. A can of beans costs $0.29. A bag of lentils costs $0.42. Estimate which is cheaper: to buy eight cans of beans or to buy five bags of lentils.

d. Jackie needs to buy 8 feet of string for each of the 28 students in the craft class.
The string costs $0.22 per foot. Estimate the total cost.

Multiply in Columns—the Easy Way

$\begin{array}{r} 38 \\ \times\ 6 \\ \hline \end{array}$	$\begin{array}{r} 38 \\ \times\ 6 \\ \hline 48 \end{array}$	$\begin{array}{r} 38 \\ \times\ 6 \\ \hline 48 \\ 180 \end{array}$	$\begin{array}{r} 38 \\ \times\ 6 \\ \hline 48 \\ +\ 180 \\ \hline 228 \end{array}$
Let's multiply 6 × 38 in parts, writing one number under the other.	First multiply 6 × 8.	Then multiply 6 × 30 and write the result under the 48. Remember, the "3" is in the tens place in the number 38 so it actually means 30.	Lastly, add.

Multiply 9 × 2.	Then 9 × 80.	Add.	Multiply 3 × 7.	Then 3 × 40.	Add.
$\begin{array}{r} 82 \\ \times\ 9 \\ \hline 18 \end{array}$	$\begin{array}{r} 82 \\ \times\ 9 \\ \hline 18 \\ 720 \end{array}$	$\begin{array}{r} 82 \\ \times\ 9 \\ \hline 18 \\ +720 \\ \hline 738 \end{array}$	$\begin{array}{r} 47 \\ \times\ 3 \\ \hline 21 \end{array}$	$\begin{array}{r} 47 \\ \times\ 3 \\ \hline 21 \\ 120 \end{array}$	$\begin{array}{r} 47 \\ \times\ 3 \\ \hline 21 \\ +120 \\ \hline 141 \end{array}$

1. Multiply.

a.

$\begin{array}{r} 76 \\ \times\ 6 \\ \hline \\ + \\ \hline \end{array}$

b.

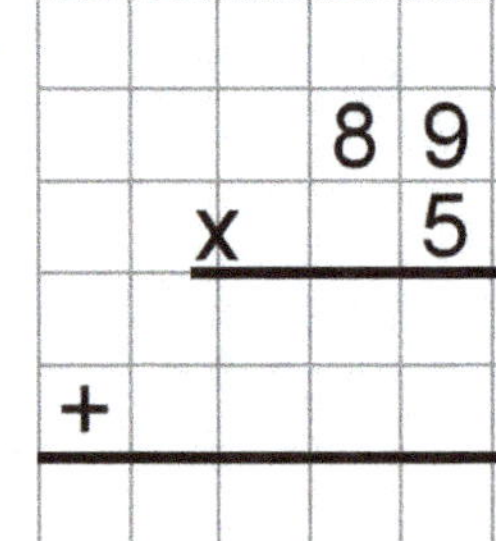

c.

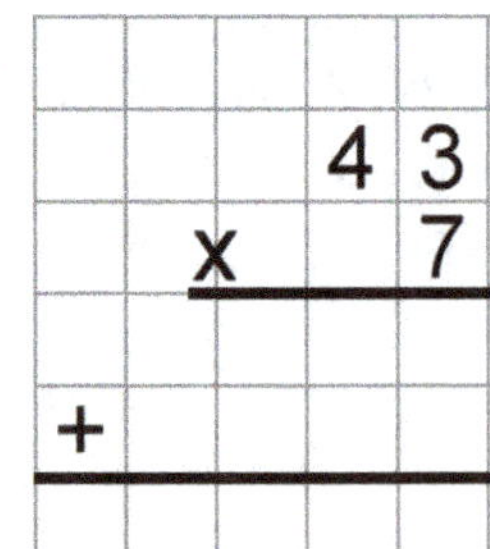

d.

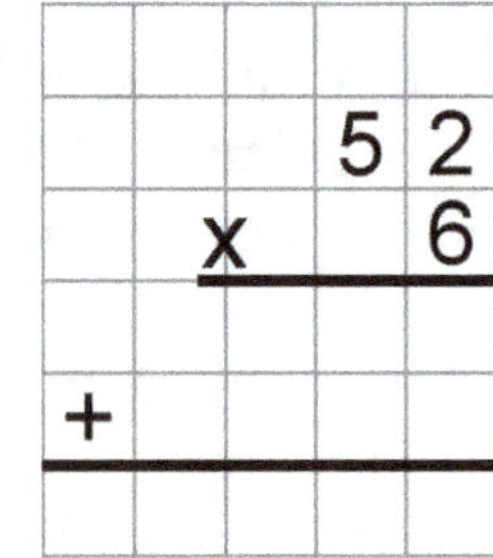

e.

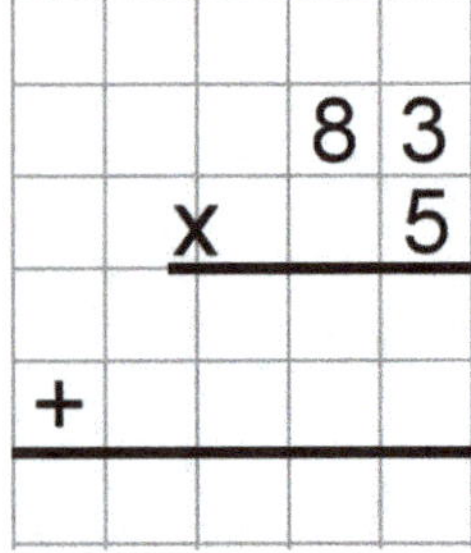

f.

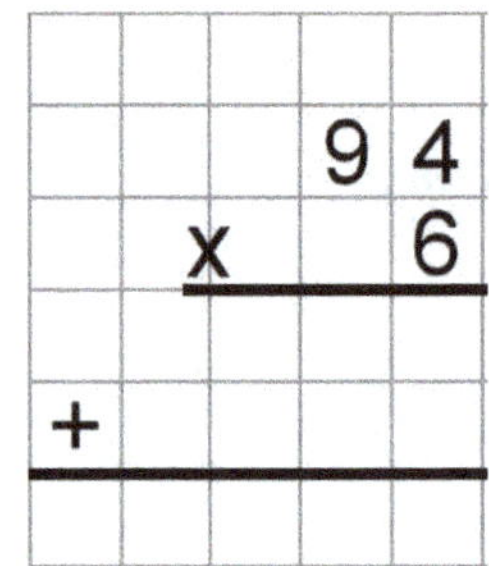

g.

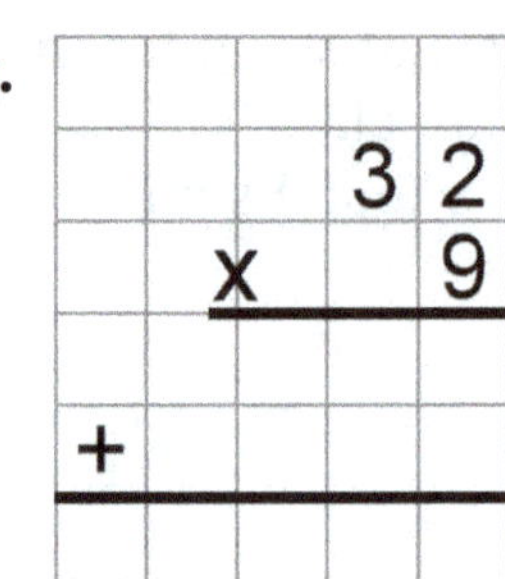

h.

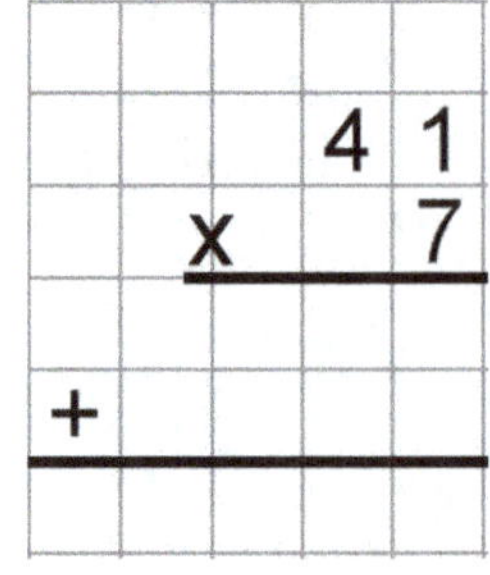

Multiplying a 3-digit number happens in exactly the same way. You multiply in parts: first the ones, then the tens, then the hundreds. Lastly, add. Just don't forget that you are multiplying *whole tens* and *whole hundreds*, not just "plain" numbers.	ones: 7 × 6 5 2 **6** × **7** ——— **4 2**	tens: 7 × 20 5 **2** 6 × **7** ——— 4 2 **1 4 0**	hundreds: 7 × 500 **5** 2 6 × **7** ——— 4 2 1 4 0 **3 5 0 0**	Add. 5 2 6 × 7 ——— 4 2 1 4 0 + 3 5 0 0 ——— 3 6 8 2

2. Multiply.

a.

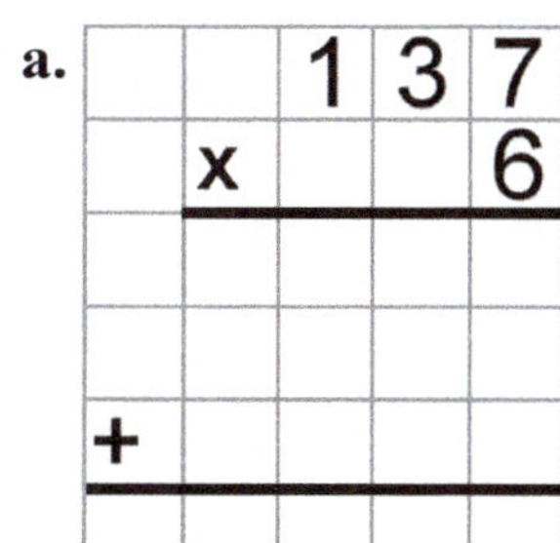

b.

$$\begin{array}{r} 218 \\ \times \quad 4 \\ \hline \\ + \\ \hline \end{array}$$

c.

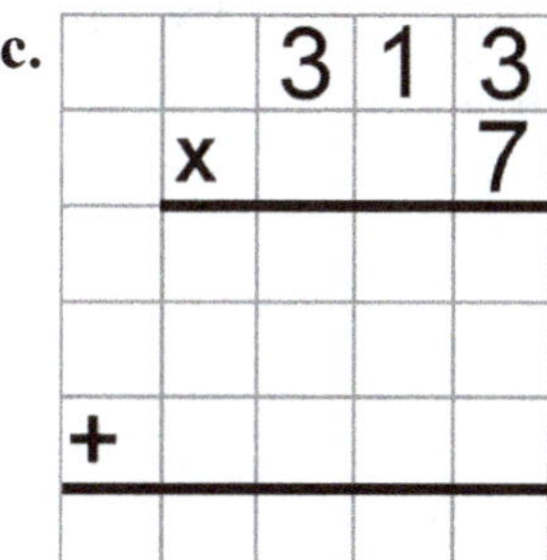

d.

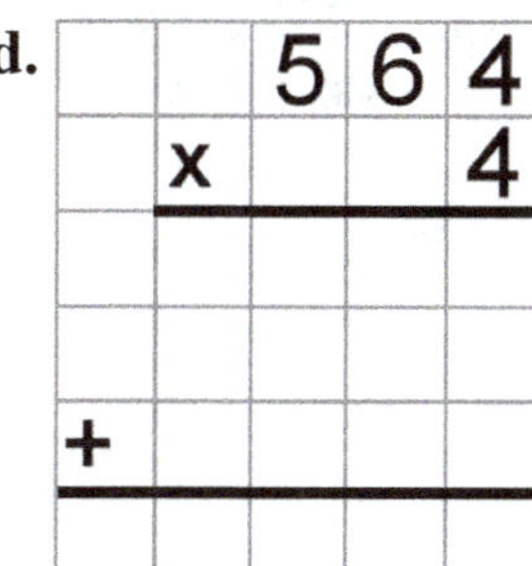

e.

f.

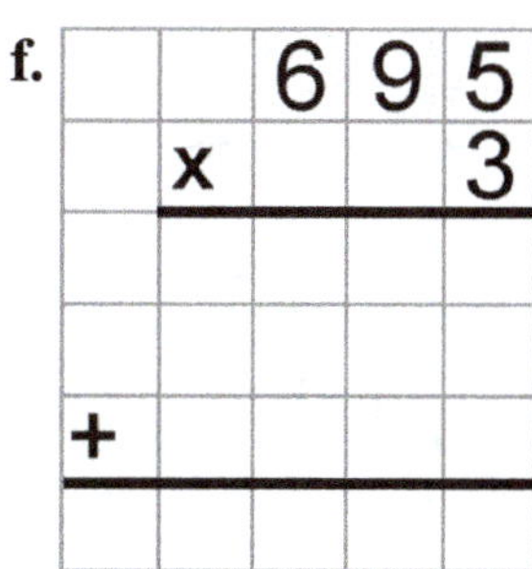

g.

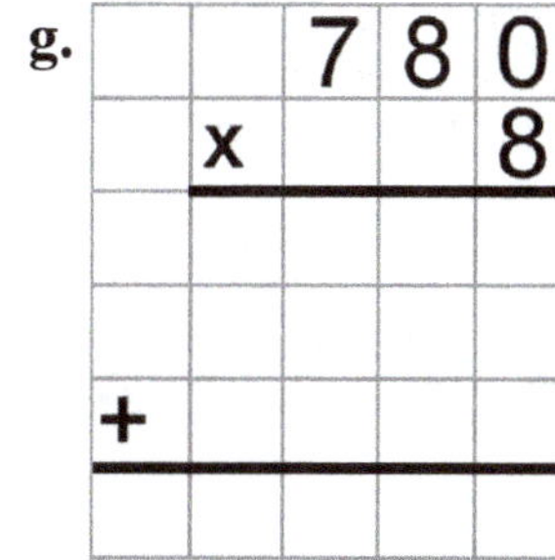

h.

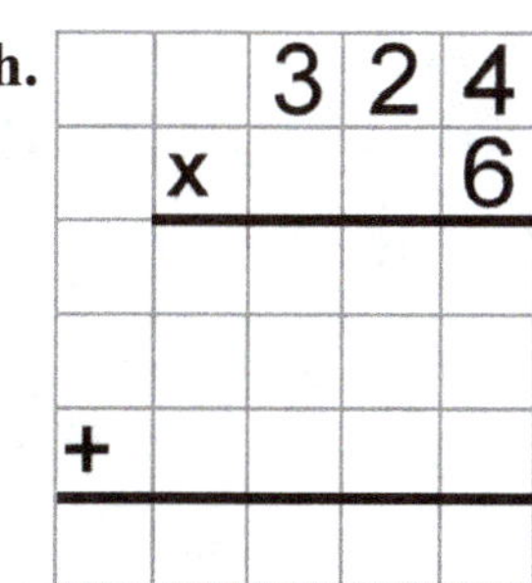

i.

$$\begin{array}{r} 927 \\ \times \quad 6 \\ \hline \\ + \\ \hline \end{array}$$

j.

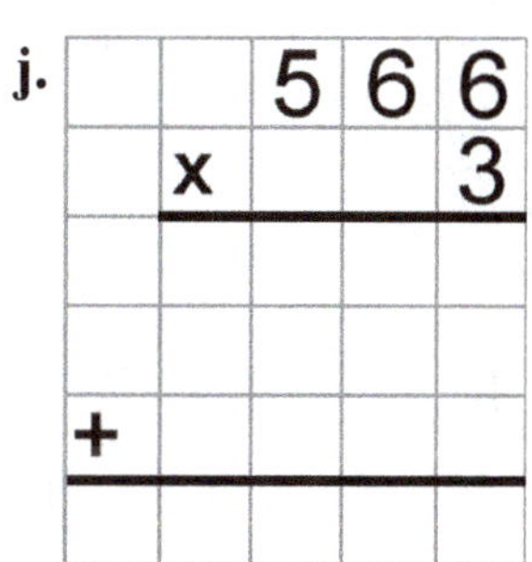

k.

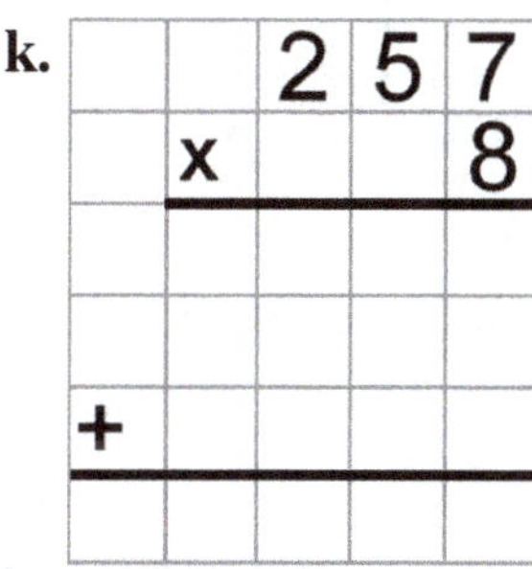

l.

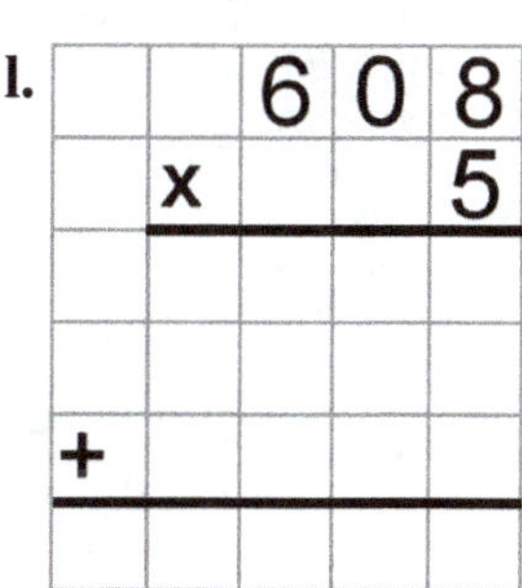

3. Solve the equations.

a. △ × 80 = 480 △ = ________	**b.** 5 × __?__ = 450 __?__ = ________	**c.** 900 × z = 81,000 z = ________

4. Solve.

a. $58 \times 5 + 291$

b. $1{,}000 - 3 \times 145$

5. Solve.

a. A $236 burglar alarm is discounted by $40. A store buys seven. What is the total cost?

b. One side of a square is 248 feet. What is its perimeter?

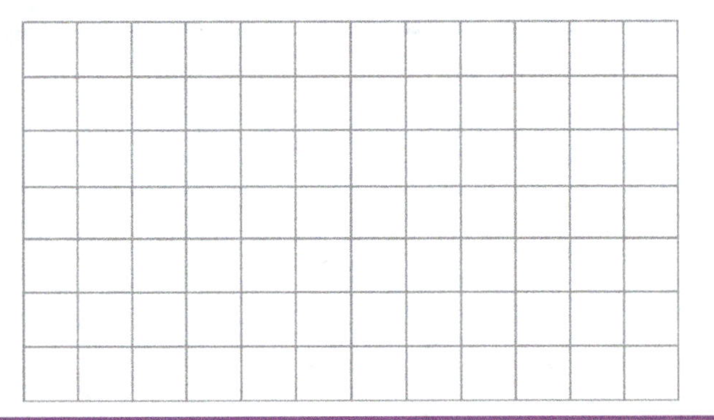

6. Solve the equations.

a. $50 \times$ __?__ $= 2{,}000$ __?__ = ________	**b.** △ $\times 60 = 2{,}400$ △ = ________	**c.** $70 \times z = 49{,}000$ z = ________
d. △ $\times 30 = 9 \times 40$ △ = ________	**e.** $5 \times$ __?__ $\times 2 = 1{,}000$ __?__ = ________	**f.** $40 \times p = 800 \times 4$ p = ________

Puzzle Corner

Figure out the missing numbers.

a.

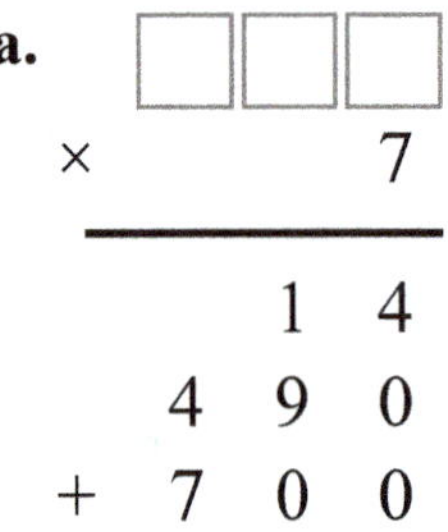

```
    □ □ □
  ×     7
  -------
      1 4
    4 9 0
  + 7 0 0
  -------
```

b.

```
      □ □ □
  ×       4
  ---------
        □ □
      2 0 0
  + 1 2 0 0
  ---------
    1 4 3 2
```

c.

```
      □ □ □
  ×       9
  ---------
        8 1
      □ □
  + 6 3 0 0
  ---------
    6 3 8 1
```

Multiply in Columns—the Easy Way, Part 2

Multiplying a 4-digit number is done in exactly the same way. You multiply in parts: first the ones, then the tens, then the hundreds, and then the thousands. Lastly, add.

Remember that you are multiplying *whole tens, whole hundreds*, and *whole thousands*—not just "plain" numbers.

ones: 7 × 8	tens: 7 × 70	hundreds: 7 × 400	thousands: 7 × 2,000	Add.
2 4 7 **8** × **7** **5 6**	2 4 **7** 8 × **7** 5 6 **4 9 0**	2 **4** 7 8 × **7** 5 6 4 9 0 **2 8 0 0**	**2** 4 7 8 × **7** 5 6 4 9 0 2 8 0 0 **1 4 0 0 0**	2 4 7 8 × 7 5 6 4 9 0 2 8 0 0 + 1 4 0 0 0 1 7 3 4 6

1. Multiply.

a.

		2	6	5	7
	x				4
			2	0	0
		2	4	0	0
+					

b.

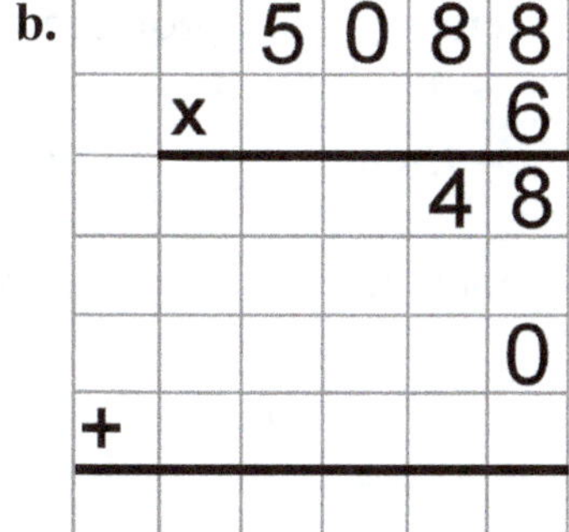

c.

		8	1	0	6
	x				3
+					

d.

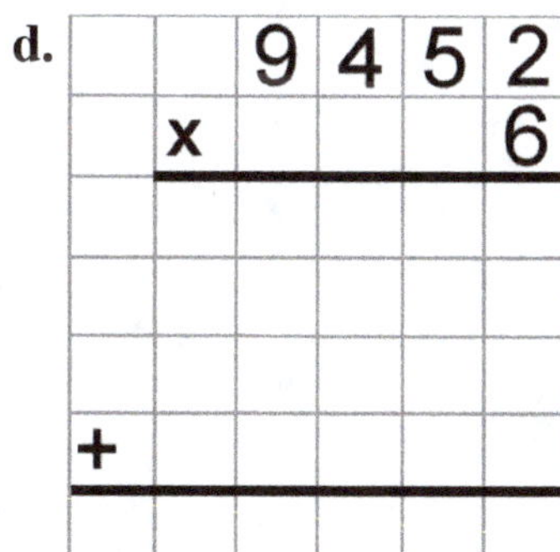

e.

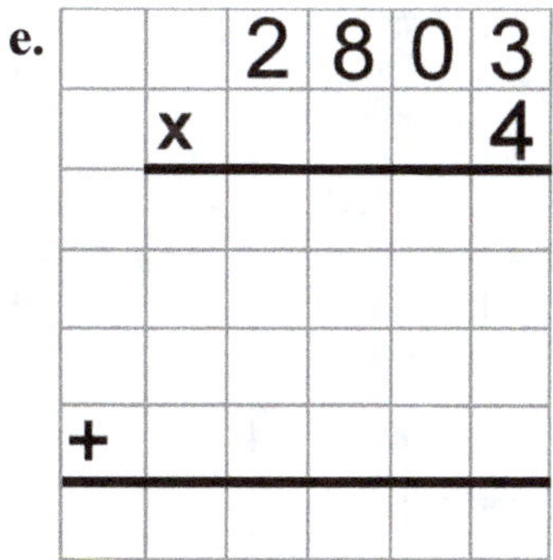

f.

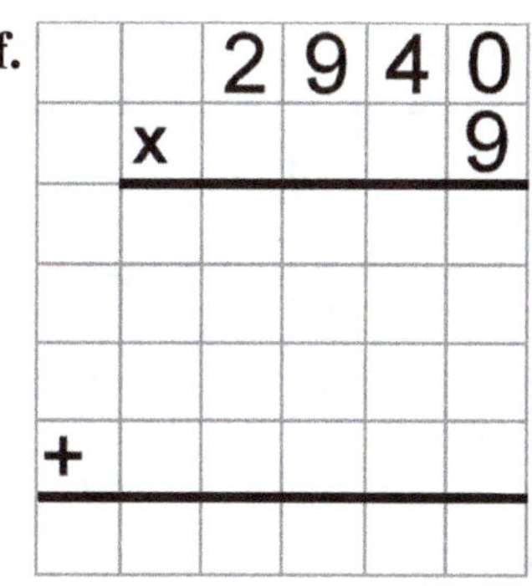

2. Multiply.

a. 5 × 1,986

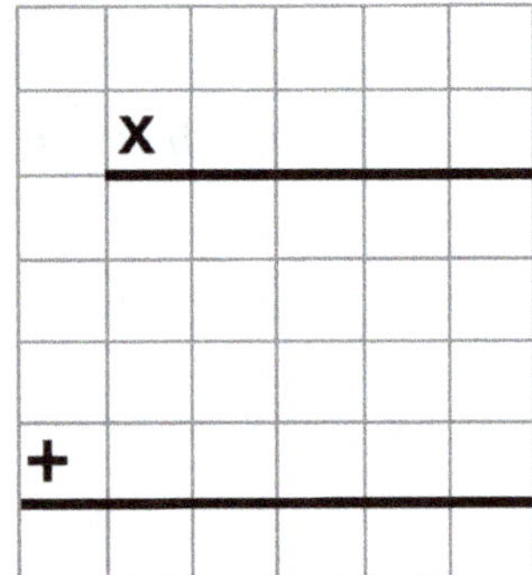

b. 9 × 2,055

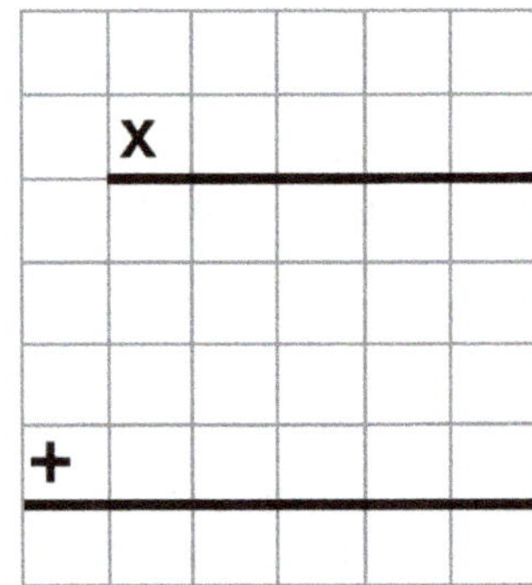

c. 7 × 3,280

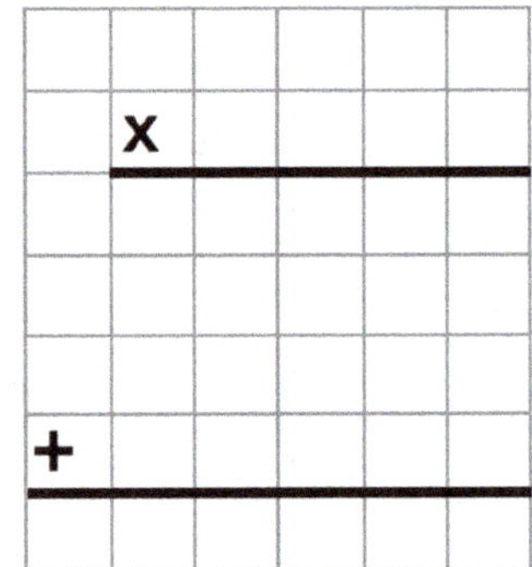

d. 6 × 6,117

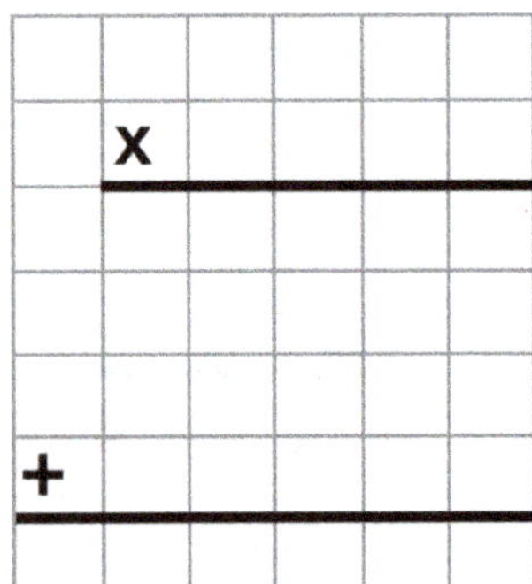

e. 5 × 2,489

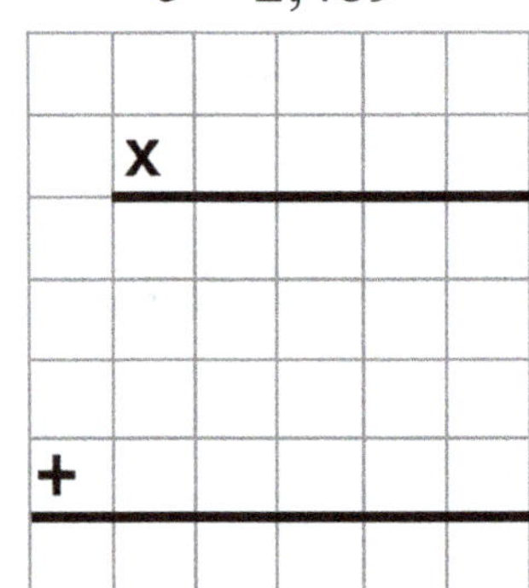

f. 8 × 4,301

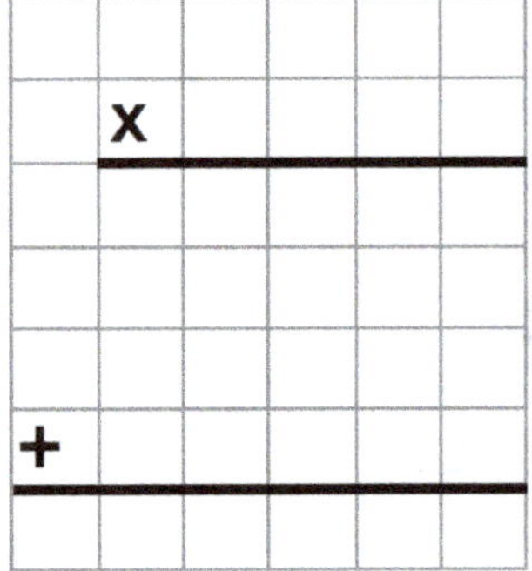

3. Solve. Write a number sentence or several for each problem.

a. The distance from Greg's house to his friend Jamie's house is 3,820 feet. What distance does Greg walk if he goes from his home to Jamie's and back *twice*?

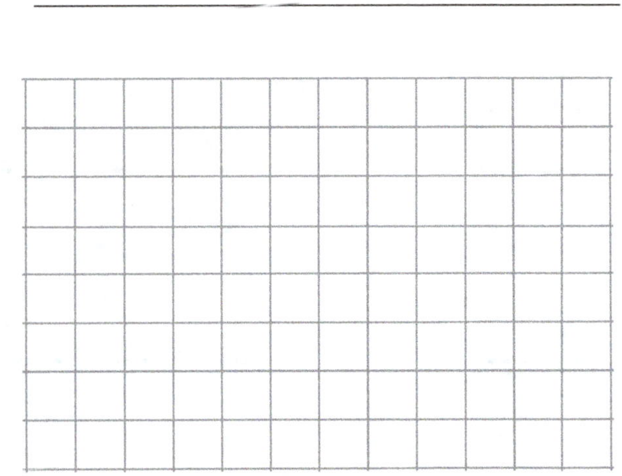

b. Jennie has 55 marbles and Sue has four times as many. How many marbles do the girls have together?

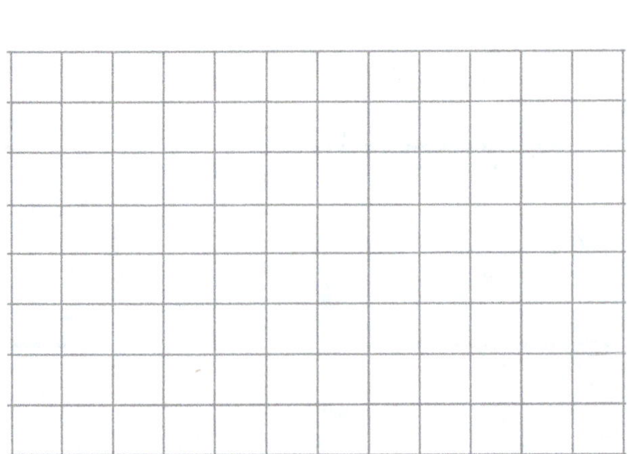

You can also multiply money amounts in parts. Multiply first the individual cents, then the ten-cent amounts, and then the whole dollars.		$5.18 × 4		$23.57 × 3
	4 × 0.08 (cents) →	0.32	3 × 0.07 →	0.21
	4 × 0.10 (ten-cents) →	0.40	3 × 0.50 →	1.50
	4 × $5 (dollars) →	+ 20.00	3 × $23 →	+ 69.00
		$20.72		$70.71

4. Multiply.

a. $1.83 × 3

b. $8.87 × 4

c. $11.50 × 7

d. $10.25 × 6

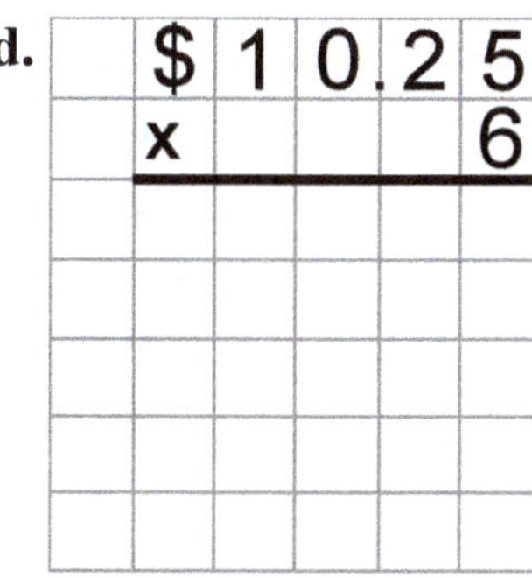

e. $6.70 × 9

f. $8.42 × 7

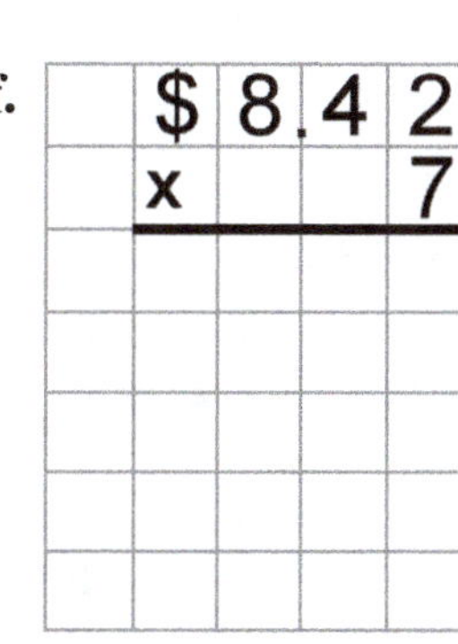

g. $20.65 × 4

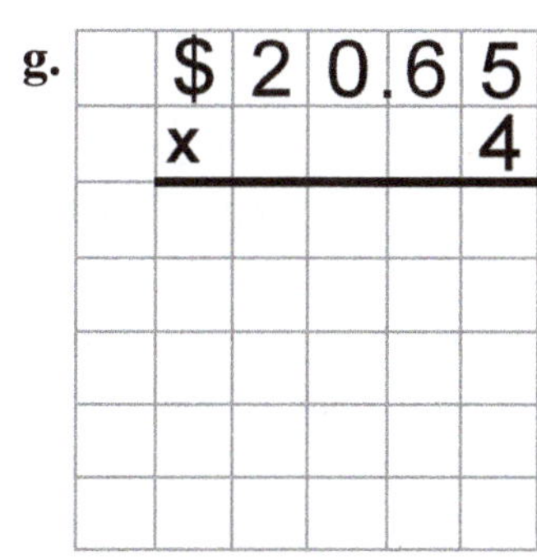

h. $53.08 × 6

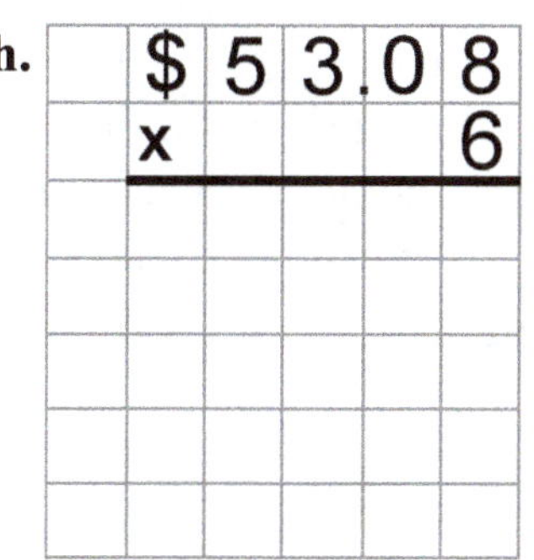

5. Solve. Write a number sentence to show your estimation on the empty line.

a. You bought nine notebooks for $1.57 each. What was the total cost?

Estimate: ______________________

b. How much is your change, if you buy eight sandwiches for $2.28 each, and pay with $20?

Estimate: ______________________

Multiplying in Columns, the Standard Way

The standard algorithm of multiplication is based on a principle you already know: multiplying in parts (partial products). We simply multiply ones, tens, and hundreds of the number separately, and then add.

However, in the standard algorithm, the additions are done *at the same time* as the multiplications – not afterwards. That way, the calculation looks more compact and takes less space.

The standard way to multiply		"The easy way"
1 6 3 × 4 ——— 2 Multiply the ones: 4 × 3 = 12. Place 2 in the ones place, but write the tens digit (1) above the tens column as a little memory note. You are *regrouping* (carrying).	1 6 3 × 4 ——— 2 5 2 Then multiply the tens, *adding* the 1 ten that was regrouped: 4 × 6 + 1 = 25 Write 25 in front of the 2. Note: This 25 means 25 tens or 250!	6 3 × 4 ——— 1 2 + 2 4 0 ——— 2 5 2 In the "easy way," we multiply in parts, and the adding is done separately.

The standard way to multiply		"The easy way"
3 7 5 × 7 ——— 5 Multiply the ones: 7 × 5 = 35 Regroup the 3 tens.	3 7 5 × 7 ——— 5 2 5 Multiply & add the tens: 7 × 7 + 3 = 52	7 5 × 7 ——— 3 5 + 4 9 0 ——— 5 2 5

1. Multiply using both methods: the standard way and the easy way.

a.		b.	
5 3 x 8 ———	5 3 x 8 ———	8 8 x 3 ———	8 8 x 3 ———

2. Multiply using both methods: the standard way and the easy way.

a.

```
  79
x  3
```

```
  79
x  3
```

b.

```
  18
x  5
```

```
  18
x  5
```

3. Multiply. Be careful with the regrouping.

a.

```
  51
x  6
```

b.

```
  19
x  3
```

c.

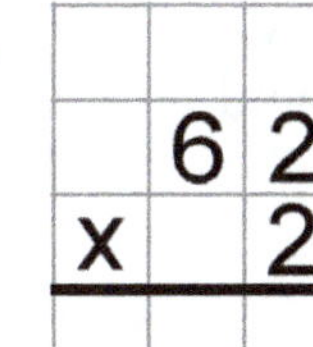

```
  62
x  2
```

d.

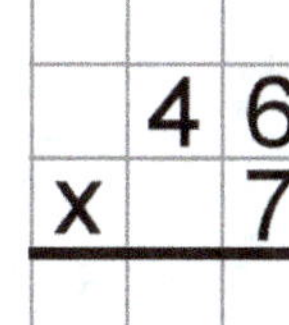

```
  46
x  7
```

e.

```
  66
x  6
```

f.

```
  39
x  9
```

g.

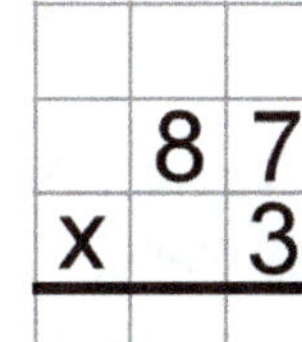

```
  87
x  3
```

h.

```
  67
x  2
```

i.

```
  20
x  9
```

j.

```
  54
x  8
```

k.

```
  34
x  6
```

l.

```
  46
x  2
```

4. Write number sentences (additions, subtractions, multiplications) on the lines, and solve.

a. What is the cost of buying three chairs for $48 each?

____

And the cost for six chairs? ____________________________

```
x
```

b. You earn $77 a day. How much do you earn in five days?

____

How much in ten days? ____________________________________

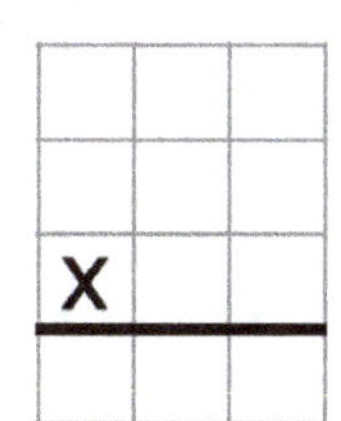

```
x
```

With a 3- or 4-digit number you have to regroup several times.

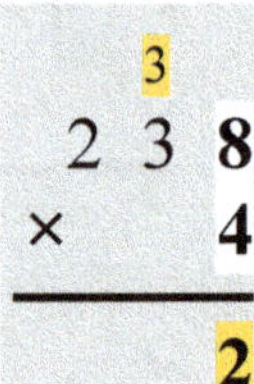

```
     3
   2 3 8
 ×     4
 -------
       2
```

Multiply the ones first.

$4 \times 8 = 32$

Write 2 in the ones place and regroup the 3 tens to the tens column.

```
   1 3
   2 3 8
 ×     4
 -------
     5 2
```

Then multiply the tens, adding the 3 regrouped tens:

$4 \times 3 + 3 = 15$

Write 5 in the tens place and regroup the 1 hundred.

```
   1 3
   2 3 8
 ×     4
 -------
   9 5 2
```

Then multiply the hundreds, adding the regrouped hundred:

$4 \times 2 + 1 = 9$

Write 9 in the hundreds place.

```
         1
   7 6 5 2
 ×       5
 ---------
         0
```

Multiply the ones:

$5 \times 2 = 10$

Write 0 in the ones place and regroup the 1 ten.

```
       2 1
   7 6 5 2
 ×       5
 ---------
       6 0
```

Then the tens. Add the regrouped ten:

$5 \times 5 + 1 = 26$

Write 6 in the tens place and regroup the 2 hundreds.

```
     3 2 1
   7 6 5 2
 ×       5
 ---------
     2 6 0
```

Multiply the hundreds:

$5 \times 6 + 2 = 32$

Write 2 in the hundreds place, and regroup the 3 thousands.

```
     3 2 1
   7 6 5 2
 ×       5
 ---------
 3 8 2 6 0
```

Multiply the thousands:

$5 \times 7 + 3 = 38$

Write 38 in front of the 260.

5. Multiply using both methods: the standard way and the easy way.

a.

```
   1 2 3
 x     8
 -------
```

```
   1 2 3
 x     8
 -------
```

b.

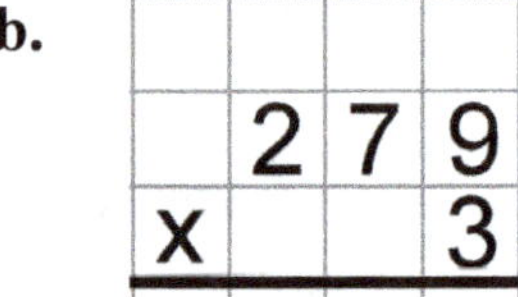

```
   2 7 9
 x     3
 -------
```

```
   2 7 9
 x     3
 -------
```

c.

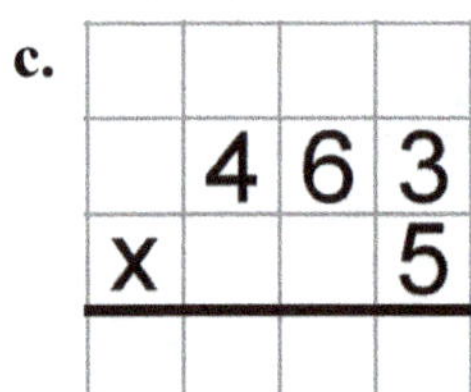

```
   4 6 3
 x     5
 -------
```

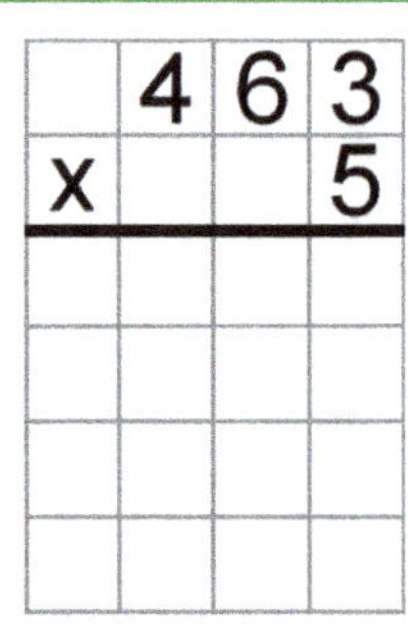

```
   4 6 3
 x     5
 -------
```

d.

```
   1 5 6
 x     6
 -------
```

```
   1 5 6
 x     6
 -------
```

6. Multiply using the standard method.

a. 462 × 2	**b.** 506 × 7	**c.** 278 × 5	**d.** 319 × 7
e. 288 × 3	**f.** 809 × 9	**g.** 287 × 3	**h.** 367 × 2
i. 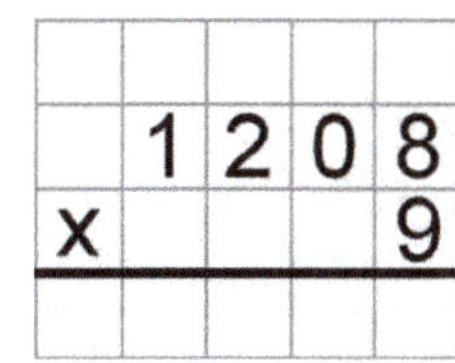 1208 × 9	**j.** 2514 × 3	**k.** 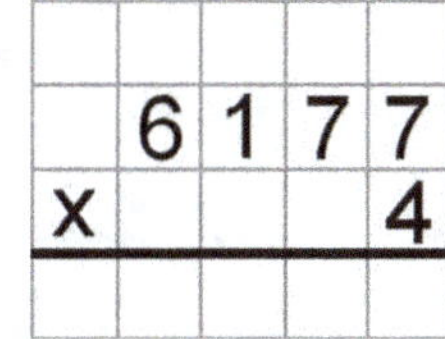 6177 × 4	**l.** 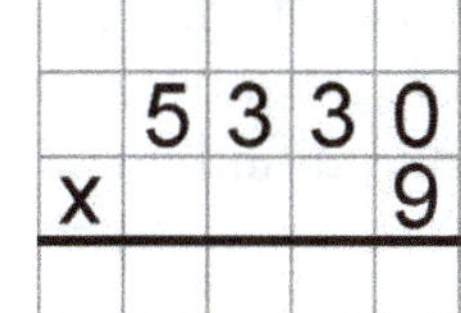 5330 × 9

7. Find the perimeter and area of a rectangular room that measures 9 ft by 28 ft.

____

____

8. Find what errors these children make.

a. Figure out what error Minnie makes every time she multiplies.

```
   4           2          4 2
   7 8         3 8        1 3 3
 ×   3       ×   9      ×     4
 -----       -----      -------
 2 5 2       2 9 7        8 1 1
```

b. This is Andy's math work. Where does he go wrong?

```
   2 8           4 5        2 1 5
 ×   3         ×   5      ×     3
 -----       -------    ---------
 6 2 4       2 0 2 5      6 3 1 5
```

Multiplying in Columns, Practice

Estimate the answer before calculating. If the estimated answer is very different from the calculated answer, there could be an error.

Estimation: $5 \times 45 \approx 5 \times 50 = 250$ The estimated answer 250 is VERY different from the calculated answer 2025. There must be an error! Can you find it and correct it? $\begin{array}{r} 4\ 5 \\ \times\ \ 5 \\ \hline 2\ 0\ 2\ 5 \end{array}$	Estimation: $7 \times 418 \approx 7 \times 400 = 2{,}800$ The estimation (2,800) and the calculated answer (2,926) are fairly close. This does *not* prove the answer is correct, but if there is an error, it is a "smaller" one. $\begin{array}{r} {\scriptstyle 1\ 5\ \ } \\ 4\ 1\ 8 \\ \times\ \ \ 7 \\ \hline 2\ 9\ 2\ 6 \end{array}$

1. First, estimate the answer. Then multiply to find the exact result.

a. Estimate: $\approx$ 9 × 40 = 360 $\begin{array}{r} 4\ 3 \\ \times\ \ 9 \\ \hline \end{array}$	**b.** Estimate: $\approx$ ___ × ______ = ______ $\begin{array}{r} 7\ 2 \\ \times\ \ 8 \\ \hline \end{array}$
c. Estimate: $\approx$ $\begin{array}{r} 7\ 7\ 1 \\ \times\ \ \ 3 \\ \hline \end{array}$	**d.** Estimate: $\approx$ ___ × ______ = ______ $\begin{array}{r} 8\ 1\ 9 \\ \times\ \ \ 5 \\ \hline \end{array}$
e. Estimate: $\approx$ $\begin{array}{r} 2\ 5\ 2\ 1 \\ \times\ \ \ \ 4 \\ \hline \end{array}$	**f.** Estimate: $\approx$ $\begin{array}{r} 8\ 7\ 1\ 2 \\ \times\ \ \ \ 3 \\ \hline \end{array}$

2. A first grade math book has 187 pages, and an eighth grade math book is three times as long. How many pages does it have?

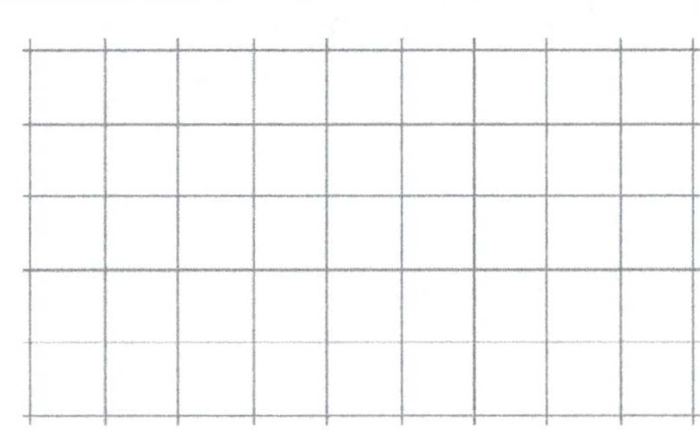

3. Solve. Write number sentence(s) to show your calculation(s). One package of paper contains 250 sheets. Marie needed 1,300 sheets. How many packages did she need to buy?

____

____

4. A school with 304 students hired buses to take the students to a museum. Each bus could seat 43 passengers.

   a. How many students could four buses seat?

   b. How many students could seven buses seat?

   c. How many buses do they need for all 304 students?

5. Solve the equations.

a.	**b.**	**c.**
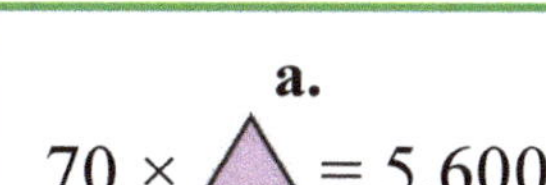 70 × △ = 5,600	60 × __?__ = 2 × 90	5 × 2 × y = 300
△ = ________	__?__ = ________	y = ________

6. Let your teacher decide if you need more practice. Multiply. Estimate the answer on the line.

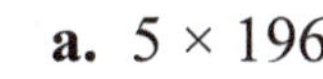 **a.** 5 × 196	**b.** 9 × 205	**c.** 9 × 9,807	**d.** 6 × 4,810
 ≈ ________	≈ ________	≈ ________	≈ ________
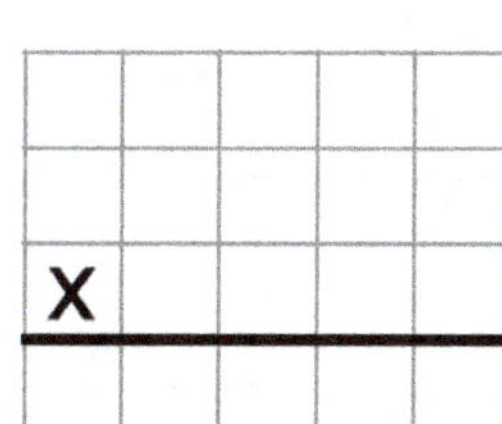	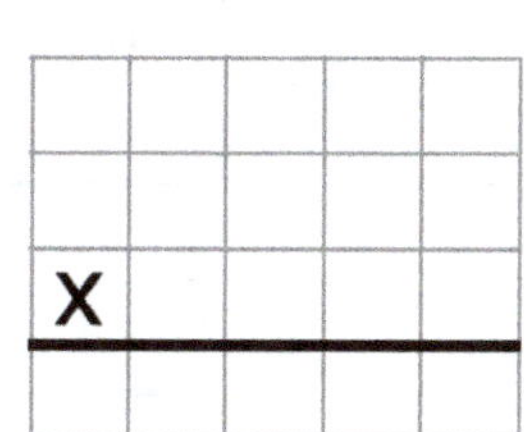	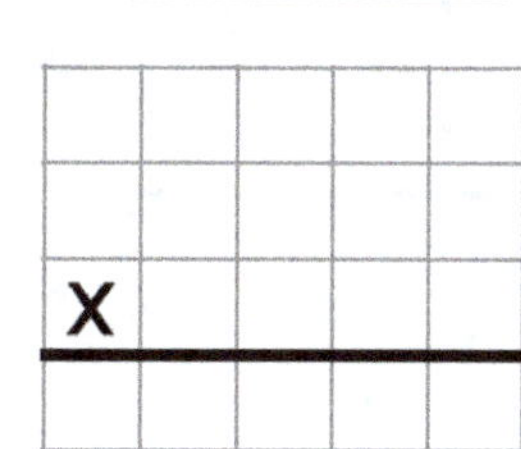	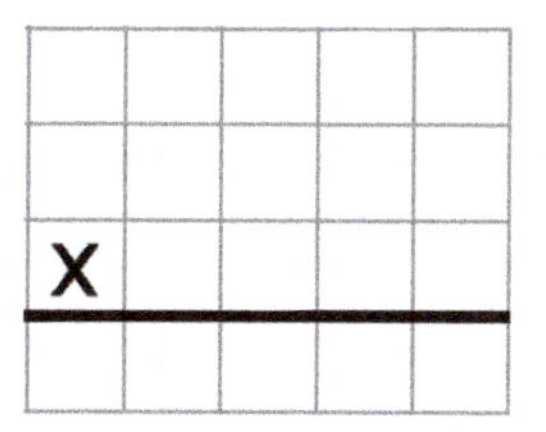

Puzzle Corner Find the missing numbers in these multiplications:

[ ] 1 [ ]	1 [ ] 4	[ ] 3 [ ]	3 2 [ ] 9
× 4	× [ ]	× 7	× 3
4 6 8	8 7 0	9 [ ] 6	[ ] 6 5 [ ]

Order of Operations Again

1. Calculate anything within parentheses ( ). 2. Do multiplications and divisions from left to right. 3. Do additions and subtractions from left to right.	$20 - 2 \times 5 + 9$ $= 20 - 10 + 9$ $= 10 + 9 = 19$	$(20 - 2) \times 5 + 9$ $= 18 \times 5 + 9$ $= 90 + 9 = 99$

1. The following calculation has five operations. In which order are they done?

$$650 - 9 \times (23 + 31) + 211$$

a. First calculate the sum _____ + ______.

b. Next multiply that sum by _____.

c. Subtract that result from _______.

d. Lastly add _____ to the subtraction result.

2. Multiply in any order. Try to find the easiest order.

a.	b.	c.	d.
$2 \times 3 \times 300 =$	$6 \times 5 \times 7 =$	$40 \times 10 \times 8 =$	$20 \times 70 \times 2 =$
$10 \times 5 \times 3 =$	$5 \times 8 \times 50 =$	$10 \times 0 \times 40 =$	$70 \times 4 \times 20 =$

3. Solve and compare the problems in each box.

a.	b.	c.
$500 - 2 \times 200 =$ _______	$70 \times (30 + 20) =$ ________	$(500 - 200) \times 2 =$ _________
$500 + 2 \times 200 =$ _______	$70 \times (30 - 20) =$ ________	$(500 + 200) \times 2 =$ _________

4. Solve.

a. $70 \times 30 - 2{,}000 =$ ________	**b.** $8 \times 200 - 200 - 500 =$ _________
c. $10 \times 7 \times (50 + 30) + 200 =$ _______	**d.** $90 + (15 + 5) \times 7 =$_________
e. $800 - 2 \times 20 + 100 =$ _________	**f.** $(500 - 50 - 50) \times 7 - 100 =$ ________

5. Calculate in the right order.

a. $5 \times 98 - 2 \times 87$

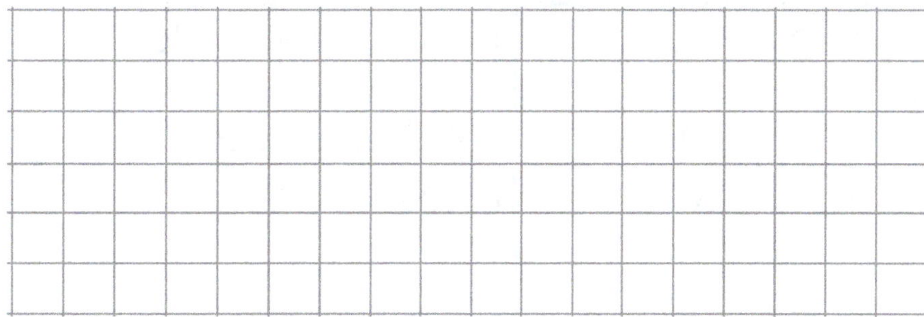

b. $2,819 - 4 \times (28 + 138)$

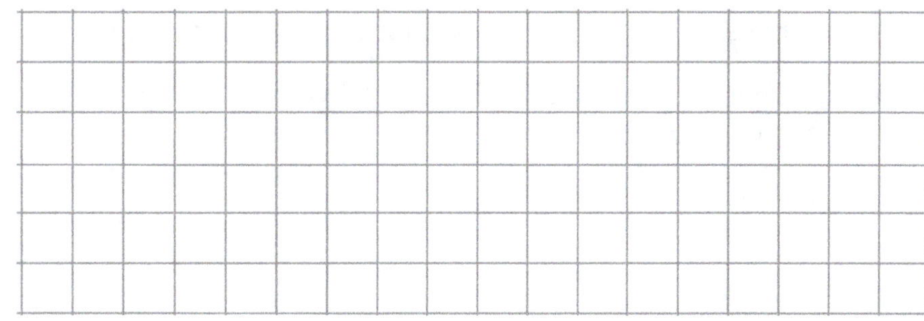

c. $8 \times (281 - 133) - 4 \times 15$

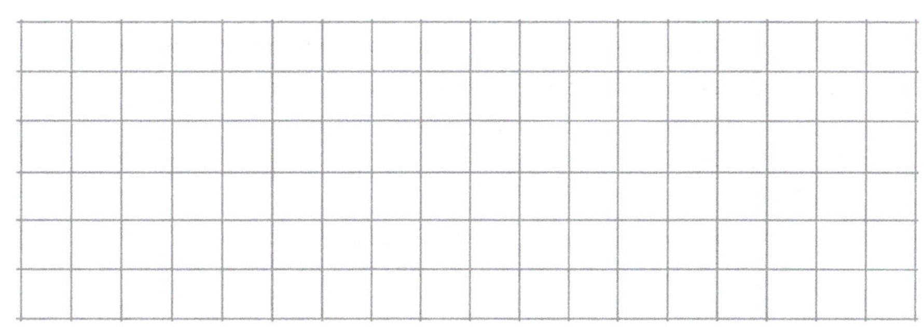

6. Find a matching number sentence for each problem, and solve.

a. Lisa bought four cards for \$2 each and three shirts for \$3 each. What was the total cost?

b. Mom bought crayons for \$2 and a book for \$3 for each of the four children. What was the total cost?

c. Mark bought five packs of paper clips for \$3 each and five pens for \$2 each. What was his change from \$50?

$4 \times (\$2 + \$3)$
$4 \times \$2 + 3 \times \$3$
$4 \times \$3 + \$2$
$4 \times \$3 \times 2$
$\$50 - 5 \times \$3 + \$2$
$\$50 - 5 \times \$3 \times 2$
$\$50 - 5 \times \$3 - 5 \times \$2$

7. Solve for the unknown N.

a. $3 \times N \times 3 = 27$ N = ________	**b.** $N \times 2 \times 100 = 2{,}400$ N = ________	**c.** $20 \times 3 \times N = 180$ N = ________
d. $N \div 5 = 300$ N = ________	**e.** $180 \div N = 20$ N = ________	**f.** $32 \div N = 16 \div 2$ N = ________

8. Write a number sentence or several for each problem and solve.

a. Elisa bought seven flash drives for $25 each.
What was her change from $200?

____

____

b. What is the total weight of eight 3-kg bags of strawberries and fifteen 2-kg bags of blueberries?

____

____

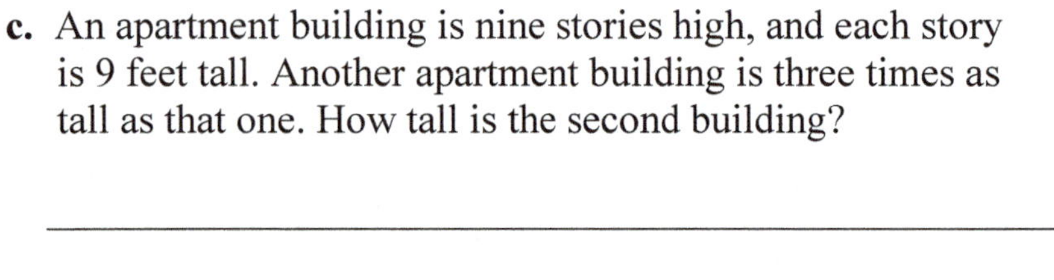

c. An apartment building is nine stories high, and each story is 9 feet tall. Another apartment building is three times as tall as that one. How tall is the second building?

____

____

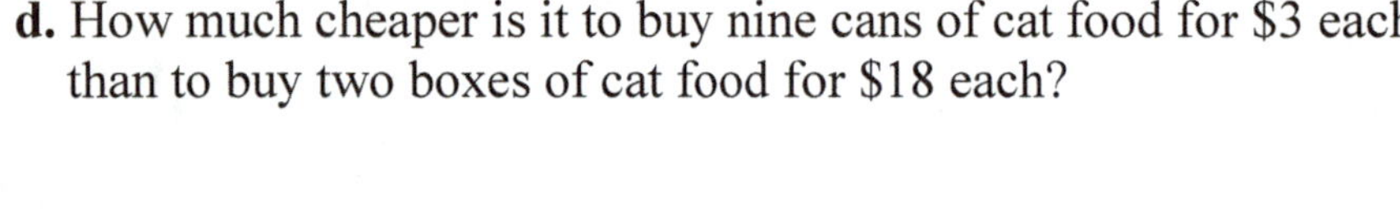

d. How much cheaper is it to buy nine cans of cat food for $3 each than to buy two boxes of cat food for $18 each?

____

____

Put operation symbols +, −, or × into the boxes and add parentheses ( ) so that the calculations become true.

a. 7 ☐ 2 ☐ 8 = 70 **b.** 80 ☐ 5 ☐ 10 ☐ 5 = 55 **c.** 4 ☐ 8 ☐ 5 ☐ 20 = 40

Money and Change

Multiplying money amounts is done the same way as multiplying whole numbers. Just remember that **the answer needs a decimal point and a dollar sign** ($). And first, estimate the result.	Estimation: 8 × $3.59 ≈ 8 × $4 = $32	$\begin{array}{r} {}^{4\ 7} \\ \$3.59 \\ \times\ \ 8 \\ \hline \$28.72 \end{array}$

1. Multiply money amounts. Estimate first.

a. Estimation: ≈ 4 × 4.50 = $18 $\begin{array}{r} \$4.55 \\ \times\ \ 4 \\ \hline \end{array}$	**b.** Estimation: ≈ ____ × ______ = ______ $\begin{array}{r} \$9.70 \\ \times\ \ 4 \\ \hline \end{array}$	**c.** Estimation: ≈ ____ × ______ = ______ $\begin{array}{r} \$4.91 \\ \times\ \ 7 \\ \hline \end{array}$
d. Estimation: ≈ ____ × ______ = ______ $\begin{array}{r} \$0.82 \\ \times\ \ 6 \\ \hline \end{array}$	**e.** Estimation: ≈ ____ × ______ = ______ $\begin{array}{r} \$12.53 \\ \times\ \ 7 \\ \hline \end{array}$	**f.** Estimation: ≈ ____ × ______ = ______ $\begin{array}{r} \$43.15 \\ \times\ \ 5 \\ \hline \end{array}$

2. If you buy eight cans of fish for $1.39 each, what is your change from $20?

Estimation:

Calculations:

3. Fill in the missing parts in the bar models and in the number sentences. Solve.

a. Jill bought three baskets for $7.20 each. She paid with $50. What was her change (x)?

$7.20	$7.20	$7.20	x

(total: $50)

$3 \times \$7.20 + x = \$50$

$x = \$\_\_\_\_\_\_\_\_ - \_\_\_\_\_ \times \$\_\_\_\_\_\_\_\_$

b. James bought four wheels for $29 each, and afterwards he had $51 left. How much did he have originally (x)?

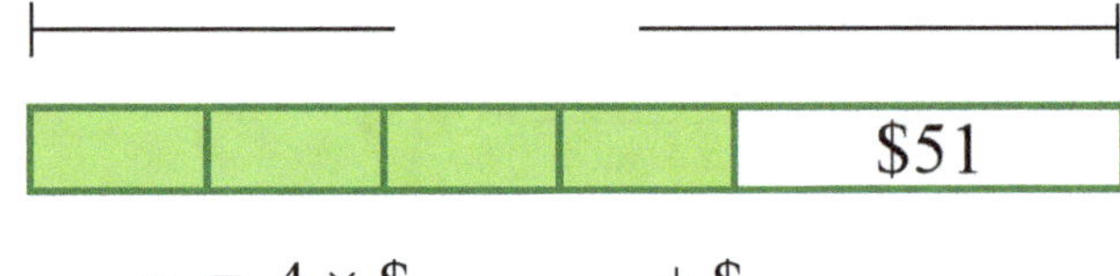

$x = 4 \times \$\_\_\_\_\_\_\_\_ + \$\_\_\_\_\_\_\_\_$

c. Jack had $30. Then he bought five screwdrivers for $3.08 apiece. How much did he have left (x)?

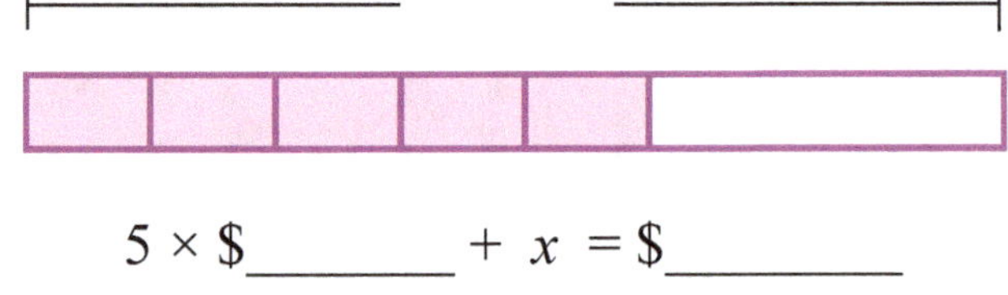

$5 \times \$\_\_\_\_\_\_\_\_ + x = \$\_\_\_\_\_\_\_\_$

$x = \$\_\_\_\_\_\_\_\_ - \_\_\_\_\_ \times \$\_\_\_\_\_\_\_\_$

d. Mom bought five meals for $11.50 per meal. Now she has $12.50 left. How much did she have originally (x)?

$x = 5 \times \$\_\_\_\_\_\_\_\_ + \$\_\_\_\_\_\_\_\_$

4. A pack of four bottles of juice costs $2.76.
   You want 20 bottles. What is the total cost?

   My estimate:

   My calculations:

5. A teacher bought 20 pencils for $0.15 each
   and 10 notebooks for $1.09 each.
   What was the total cost?

   My estimate:

   My calculations:

6. Paul bought four computer mice for $9.80 each
   and a hard drive for $65, and had $25.80 left.
   How much money did Paul have initially?

   My estimate:

   My calculations:

Puzzle Corner

A construction company bought eight wheelbarrows. The wheelbarrows used to cost $145 but now they were discounted. The total bill came to exactly $1,000. How much was the discount for one wheelbarrow?

So Many of the Same Thing

1. Fill in the tables, thinking logically!

a. A bus is traveling at the speed of 45 miles per hour. That means it travels 45 miles in one hour. Fill in the table how many miles the bus travels in the given amount of hours.

Miles	45									
Hours	1	2	3	4	5	6	7	8	9	10

b. One meter of fabric costs $5.10. Fill in the table.

Dollars	$5.10									
Meters	1	2	3	4	5	6	7	8	9	10

c. Two cans of beans cost $3.00. Fill in the table.

Dollars		$3.00								
Cans	1	2	3	4	5	6	7	8	9	10

d. You can get four buckets of paint for $60. Fill in the table.

Dollars		$60								
Buckets	2	4	6	8	10	12	14	16	18	20

e. An earthworm can travel at the speed of 240 feet per hour. Notice that the table has minutes!

Feet										
Minutes	10	20	30	40	50	60	70	80	90	100

f. Ernie can patch four bicycle tires in an hour.

Tires	Minutes
1	
2	
3	
4	
5	
6	

g. Maria can knit three scarves in nine days.

Days	Scarves
3	
6	
9	
12	
15	
18	

h. Jack earned $75 in five hours.

Hours	Dollars
1	
2	
3	
4	
5	
6	

Two identical bags contain 12 kg of potatoes in total. How many such bags would you need to have 30 kg of potatoes?

To solve this problem, you can make a chart like the one on the right:

1 bag	_____ kg
2 bags	12 kg
3 bags	3 × 6 kg = 18 kg
____ bags	____ × 6 kg = 30 kg

The total number of chocolates in five identical boxes was 30. How many chocolates would two boxes contain?

First find out how many are in ONE box:

1 box	_____ chocolates
2 boxes	_____ chocolates
5 boxes	30 chocolates

2. Solve the problems using the tables. First find out the answer for ONE of the things.

a. Six flowers cost $18. How much would five flowers cost?

1 flower	
5 flowers	
6 flowers	$18

b. Four cans of peas weigh 800 g. How much would three cans weigh?

4 cans	800 g

c. A package of three fishing lures costs $6. How much would seven lures cost?

3 lures	$6

d. Mark can watch three episodes of the "Animal Farm" series in 90 minutes. How long would it take for him to watch 5 episodes?

e. Mark did five sit-ups in ten seconds. How many could he do in one minute if he maintains the same speed?

f. Ten notebooks cost $20.
What would seven cost?

g. Ann read 120 pages in four days. At that rate, how many days would it take her to read a 300-page book?

h. Six pairs of socks cost $4.50. What do 30 pairs of socks cost?

3. These problems are more challenging.

a. Five collectible cars cost $35.50. What would four cars cost?

b. Margie can weed five rows of strawberry plants in 1 hour and 40 minutes. How long would it take her to weed nine rows?

c. Elaine can run around the track four times in an hour. Today, she only ran around the track three times, and then walked around it the fourth time. In all, this took her 10 minutes longer than her normal schedule. How many minutes did it take her to walk around the track, assuming that she ran at the same pace as usual?

Multiplying Two-Digit Numbers in Parts

The picture illustrates the multiplication 18 × 27 using an area model.

The sides of the whole rectangle are 18 and 27. It is divided into four parts. The areas of the partial rectangles are:

10 × 20 = 200 (top left)
10 × 7 = 70 (top right)
8 × 20 = 160 (bottom left)
8 × 7 = 56 (bottom right)

Of course, we add those to find the total area:

200 + 70 + 160 + 56 = 486 square units

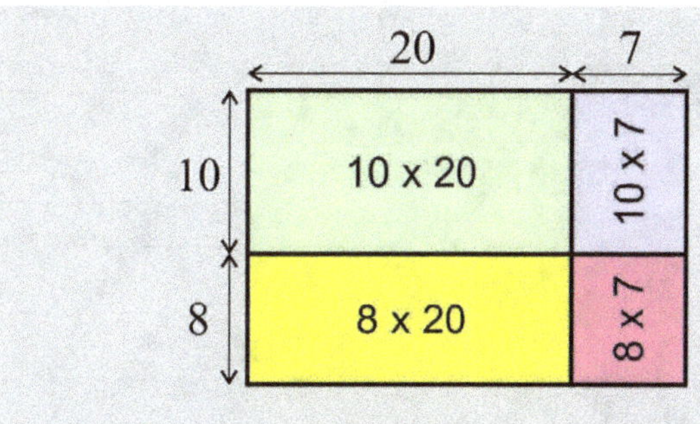

18 × 27 = 10 × 20 + 10 × 7
+ 8 × 20 + 8 × 7
= 200 + 70 + 160 + 56 = **486**

1. Fill in the missing numbers. Write the area of the *whole* rectangle as a SUM of the areas of the *smaller* rectangles. Also find the total area.

a. 23 × 31 = _____ × _____ + _____ × _____
+ _____ × _____ + _____ × _____
= _____ + _____ + _____ + _____
= ______

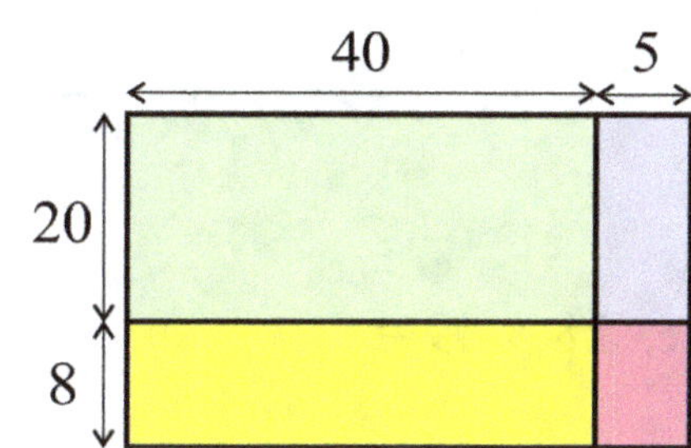

b. 28 × 45 = _____ × _____ + _____ × _____
+ _____ × _____ + _____ × _____
= _____ + _____ + _____ + _____
= ______

40
5
20
8

c. _____ × _____ = _____ × _____ + _____ × _____
+ _____ × _____ + _____ × _____
= _____ + _____ + _____ + _____
= ______

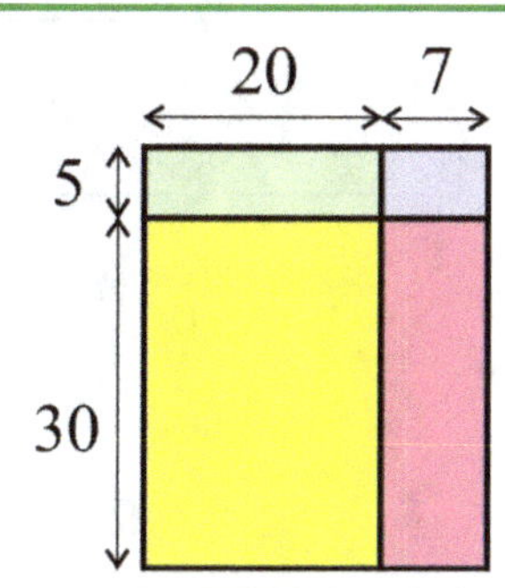

2. It is your turn to draw. Draw a four-part rectangle to illustrate the multiplications, as it is in the previous problem. You don't have to draw to scale —a sketch is good enough.

a. $13 \times 27 =$

_____ × ______ + _____ × _____

+ _____ × ______ + _____ × _____

=

b. $36 \times 25 =$

_____ × ______ + _____ × _____

+ _____ × ______ + _____ × _____

=

c. $28 \times 49 =$

_____ × ______ + _____ × _____

+ _____ × ______ + _____ × _____

=

We can also use our "easy way" of multiplying with two 2-digit numbers. The calculation will have FOUR parts, just like the area model you just studied.

Let's multiply **56 × 28** in four parts: 6 × 8, 6 × 20, 50 × 8, and 50 × 20.

Step 1	Step 2	Step 3	Step 4	Step 5
2 **8** × 5 **6** ——— **4 8**	**2** 8 × 5 **6** ——— 4 8 **1 2 0**	2 **8** × **5** 6 ——— 4 8 1 2 0 4 0 0	**2** 8 × **5** 6 ——— 4 8 1 2 0 4 0 0 1 0 0 0	**2** 8 × **5** 6 ——— 4 8 1 2 0 4 0 0 + 1 0 0 0 ——— 1 5 6 8
First multiply 6 × 8.	Then 6 × 20. Notice the "2" means 20, not just 2!	Next, 50 × 8. Notice the "5" means 50.	Then, 50 × 20. Notice the "5" means 50 and the "2" means 20!	Lastly add.

3. It is your turn to practice. Multiply the numbers in four parts, and lastly add.

a.

  8 7
× 1 5

5 × 7 →
5 × 80 →
10 × 7 →
10 × 80 → +

b.

  2 4
× 7 1

1 × 4 →
1 × 20 →
70 × 4 →
70 × 20 → +

c.

  3 8
× 9 2

___ × 8 →
___ × 30 →
90 × 8 →
90 × 30 → +

d.

  5 2
× 6 5

___ × ___ →
___ × ___ →
60 × 2 →
60 × 50 → +

4. Multiply.

a.	b.	c.	d.
$\begin{array}{r} 55 \\ \times\ 12 \\ \hline \end{array}$ + ______	$\begin{array}{r} 81 \\ \times\ 64 \\ \hline \end{array}$ + ______	$\begin{array}{r} 73 \\ \times\ 80 \\ \hline \end{array}$ + ______	$\begin{array}{r} 99 \\ \times\ 11 \\ \hline \end{array}$ + ______

5. Multiply in four parts using the "easy way." Draw an area model to illustrate each multiplication.

a.

$$\begin{array}{r} 24 \\ \times\ 17 \\ \hline \end{array}$$

+ ______

b.

$$\begin{array}{r} 44 \\ \times\ 39 \\ \hline \end{array}$$

+ ______

6. Multiply in four parts using the "easy way." Draw an area model to illustrate each multiplication.

a.

$$\begin{array}{r} 6\ 2 \\ \times\ 3\ 3 \\ \hline \end{array}$$

$+$

---

b.

$$\begin{array}{r} 4\ 7 \\ \times\ 5\ 3 \\ \hline \end{array}$$

$+$

---

c.

$$\begin{array}{r} 8\ 3 \\ \times\ 2\ 9 \\ \hline \end{array}$$

$+$

Multiply by Whole Tens in Columns

$7 \times 58 = 406$. Now, based on that, what would 70×58 be? Can you guess? $\begin{array}{r} ^{5} \\ 5\,8 \\ \times\;\;7 \\ \hline 4\,0\,6 \end{array}$	$116 \times 9 = 1{,}044$. Based on that, what would 116×90 be? Can you guess? $\begin{array}{r} ^{1\;5} \\ 1\,1\,6 \\ \times\;\;\;9 \\ \hline 1\,0\,4\,4 \end{array}$

Don't read more until you think about the questions above!

70×58 $= 10 \times 7 \times 58$ So, the result to 70×58 is ten times the result to 7×58. Since $7 \times 58 = 406$, then 70×58 is 4,060. Just tag a zero!	116×90 $= 116 \times 9 \times 10$ So, the result to 116×90 is *ten* times the result to 116×9. Since $116 \times 9 = 1{,}044$, then 116×90 is 10,440. Just tag a zero!

1. Use the above method to solve these problems.

a. $60 \times 87 =$ ____________

(First multiply 6×87)

b. $51 \times 40 =$ ____________

(First multiply $51 \times$ ____)

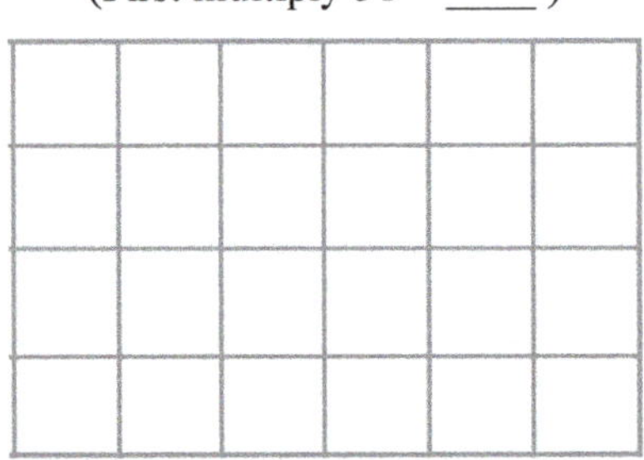

c. $66 \times 30 =$ ____________

(First multiply _____ $\times$ _____)

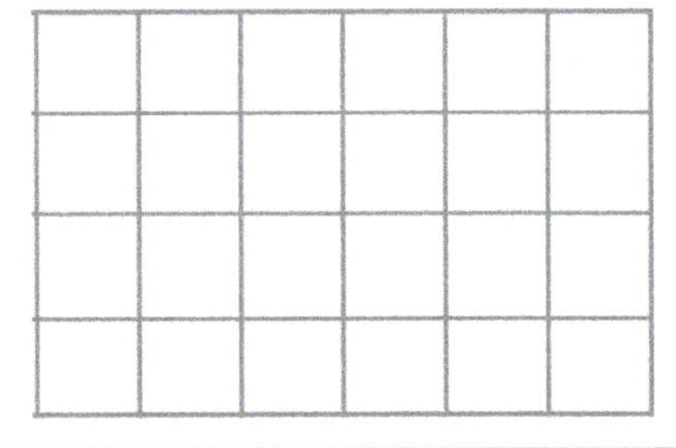

2. **a.** A crate of apples weighs 20 kg.
How much do 65 crates weigh?

b. One crate contains four layers of apples.
There are 25 apples in each layer.
How many apples are in a crate?

c. A store owner sold 60 kg of apples to one customer.
How many apples did the customer get?

$\begin{array}{r} {}^{1\ 3} \\ 5\ 2\ 8 \\ \times \quad\ 4 \\ \hline 2\ 1\ 1\ 2\ 0\ 0 \end{array}$	To multiply 5,280 × 40, you can first multiply 528 × 4 (without the ending zeros), and then write ***two* zeros after the answer**. We get 5,280 × 40 = 211,200.

3. Multiply. First multiply without the ending zeros, then add them to the answer.

a. 800 × 46 = ___________

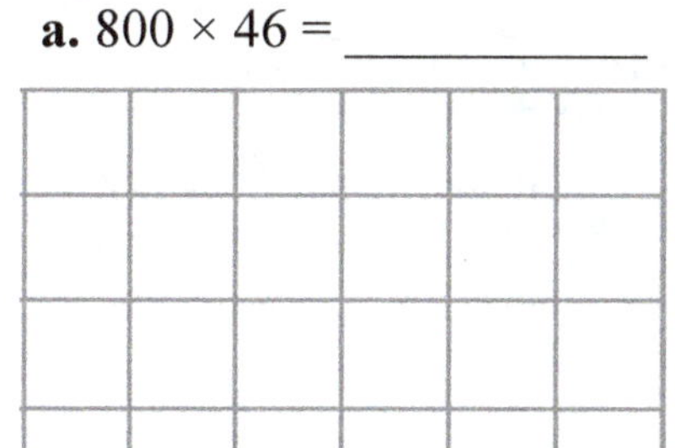

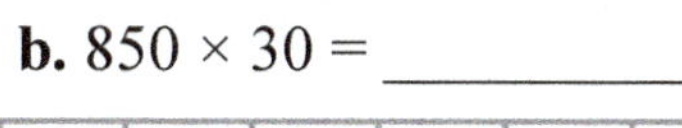

b. 850 × 30 = ___________

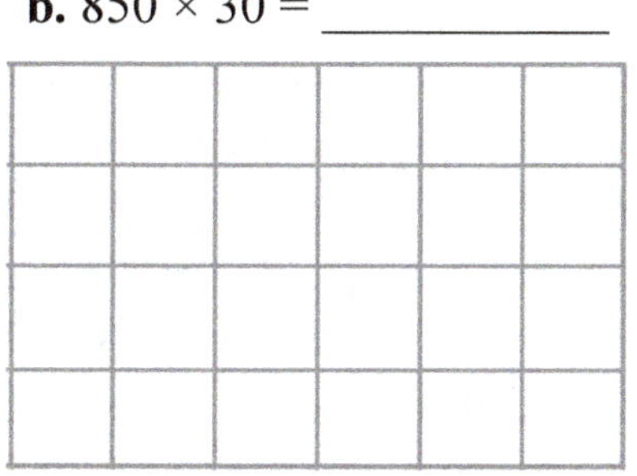

c. 109 × 40 = ___________

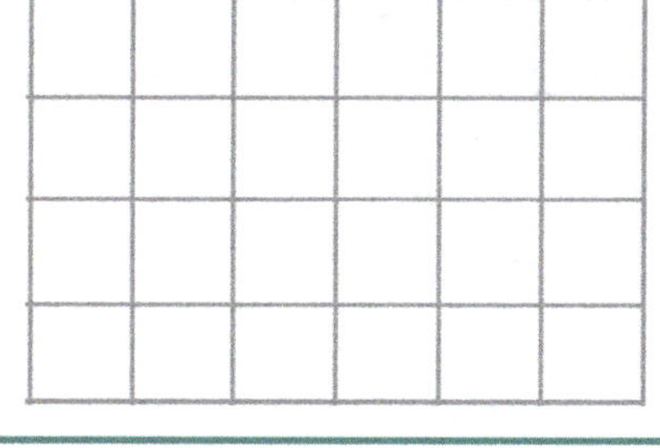

d. 120 × 70 = ___________

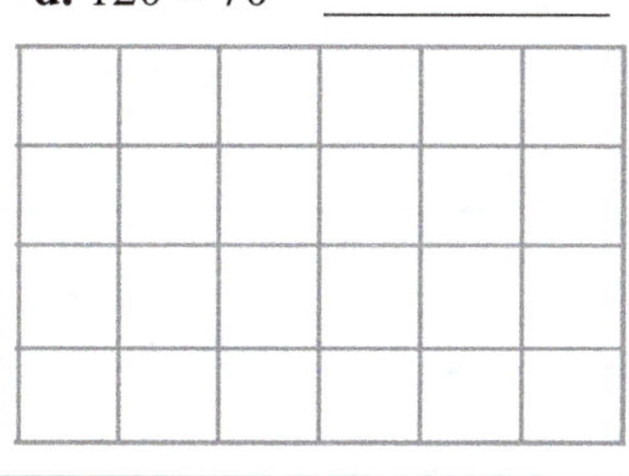

e. 400 × 335 = ___________

f. 1,620 × 40 = ___________

4. Mr. Hendrickson drives his bus about 250 km each day on his route. About how many kilometers does he drive during his 5-day work week?

   How much in the 4 weeks he works in a month?

5. *One* side of farmer Greg's square field measures 200 m. He jogged around the square seven times. What is the distance he jogged?

6. Calculate 65,000 − 50 × 430. Use a notebook.

If 7 × 8 × 9 × 10 = 5,040
then what is 14 × 16 × 45 × 50?
(Try not to actually calculate in columns! There is a quicker way.)

Multiplying in Parts: Another Way

(This lesson is optional)

You have learned to do 7 × 82 in parts: first multiply 7 × 80 and then 7 × 2. The same idea works when we have two 2-digit numbers.

Let's solve **25 × 34**. To find 25 times some number we can find 20 times the number and 5 times the number, and then add those two. So we break 25 × 34 into two parts: 20 × 34 and 5 × 34.

Example 1: 25 × 34			**Example 2: 78 × 47**		
Do 20 × 34. Don't forget the extra zero.	Then do 5 × 34.	Then add the parts.	Do 70 × 47. Remember the extra zero.	Then do 8 × 47.	Then add the parts.
$\begin{array}{r} 34 \\ \times\ 20 \\ \hline 680 \end{array}$	$\begin{array}{r} {}^{2}\ \\ 34 \\ \times\ 5 \\ \hline 170 \end{array}$	$\begin{array}{r} 680 \\ +\ 170 \\ \hline 850 \end{array}$	$\begin{array}{r} {}^{4}\ \\ 47 \\ \times\ 70 \\ \hline 3290 \end{array}$	$\begin{array}{r} {}^{5}\ \\ 47 \\ \times\ 8 \\ \hline 376 \end{array}$	$\begin{array}{r} 3290 \\ +\ 376 \\ \hline 3666 \end{array}$

1. Break the multiplications into two parts. You don't have to find the answer.

a. <u>28</u> × 16 = ***20*** × 16 and ***8*** × 16	**b.** <u>48</u> × 73 = _____ × 73 and _____ × 73
c. <u>19</u> × 42 = _____ × 42 and _____ × 42	**d.** <u>55</u> × 89 = _____ × 89 and _____ × 89

2. Find the final answers for the problems in (1).

a.

× ______ × ______ + ______

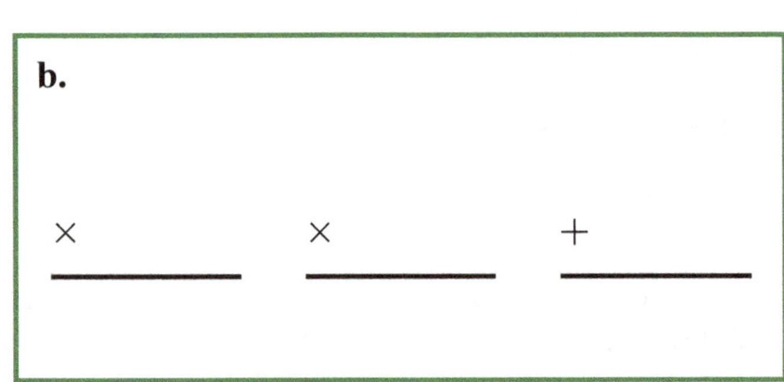

c.

× ______ × ______ + ______

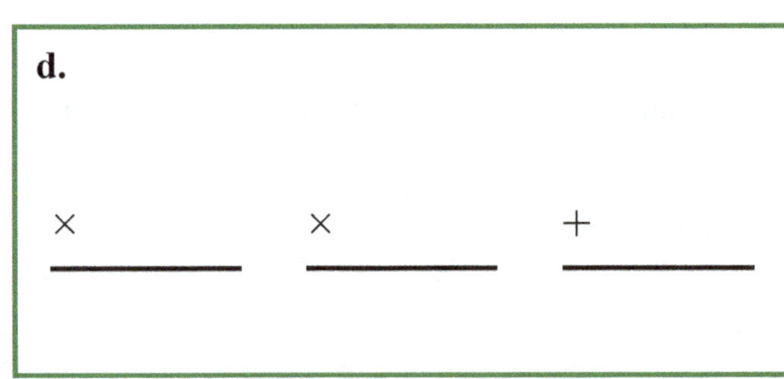

3. Break the multiplications into two parts. Then calculate to find the final answer.

a. **46** $\times$ 41 = ____ $\times$ ____ and ____ $\times$ ____

$\times$ ______ $\times$ ______ + ______

b. **28** $\times$ 39 = ____ $\times$ ____ and ____ $\times$ ____

$\times$ ______ $\times$ ______ + ______

c. 15 $\times$ 27 = ____ $\times$ ____ + ____ $\times$ ____

$\times$ ______ $\times$ ______ + ______

d. 93 $\times$ 16 = ____ $\times$ ____ + ____ $\times$ ____

$\times$ ______ $\times$ ______ + ______

Solve. Estimate before you calculate.

4. It costs Mary $27 per month for Internet service. How much does Mary pay per year for Internet?

   Estimation:

5. Find the product 1 $\times$ 2 $\times$ 3 $\times$ 4 $\times$ 5 $\times$ 6.

6. A store sells 15-kg boxes of apples for $35 a box. If you buy 12 boxes, what is their total weight?

   Estimation:

7. What is the total cost for 12 boxes of apples?

   Estimation:

The Standard Multiplication Algorithm with a Two-Digit Multiplier

You have learned to calculate multiplications such as 67 × 53 in parts. You did two multiplications and then added. It took three separate calculations.

In the traditional way of multiplying there are also three separate calculations, but all three calculations appear together. Study how we solve 67 × 53 below.

```
   2
   5 3
 × 6 7
 -----
 3 7 1
```

First multiply **7 × 53**. Pretend the 6 of the 67 is not there.

```
   1 2
   5 3
 × 6 7
 -------
   3 7 1
 3 1 8 0
```

Then multiply **60 × 53**, but write the result under the 371. Remember the zero. Pretend that the 7 of the 67 is not there. You can cross out the carry number from the previous calculation, so you won't get confused by it.

```
    1 2
    5 3
  × 6 7
 --------
    3 7 1
 +3 1 8 0
 --------
  3 5 5 1
```

Lastly add.

Study these examples, too. Note we need an extra zero in the ones place on the second line!

5 × 34

```
   2
   3 4
 × 2 5
 -----
 1 7 0
```

20 × 34

```
   3 4
 × 2 5
 -----
 1 7 0
 6 8 0
```

Add.

```
     3 4
   × 2 5
 -------
   1 7 0
 + 6 8 0
 -------
   8 5 0
```

4 × 63

```
   1
   6 3
 × 9 4
 -----
 2 5 2
```

90 × 63

```
     2
     6 3
   × 9 4
 -------
   2 5 2
 5 6 7 0
```

Add.

```
      6 3
    × 9 4
 --------
    2 5 2
 +5 6 7 0
 --------
  5 9 2 2
```

1. Fill in the missing digits and complete the calculations.

a.

			6	5
	x		1	8
+		6	5	0

b.

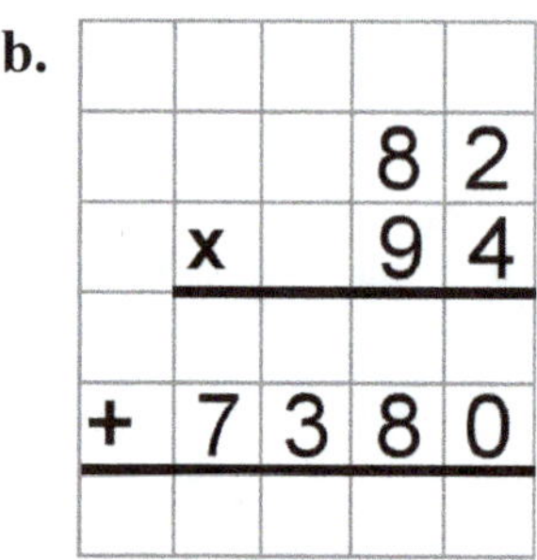

c.

			9	3
	x		2	2
		1	8	6
+				

d.

			7	0
	x		5	3
		2	1	0
+				

2. Multiply.

a.

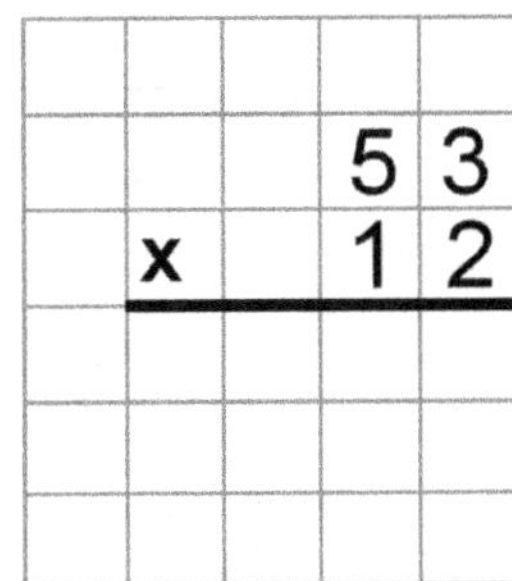

b.

11
x 35

c.

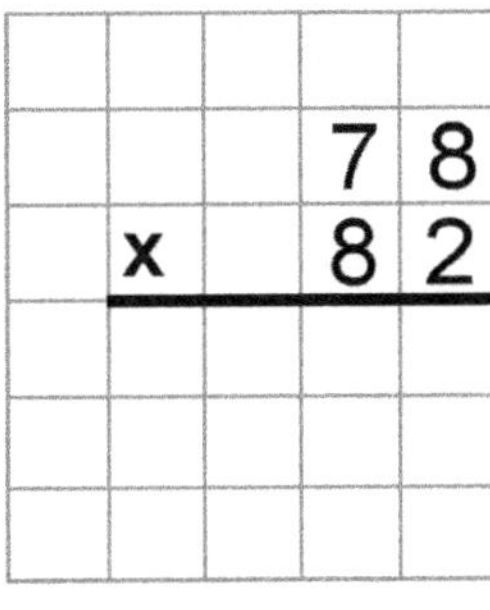

d.

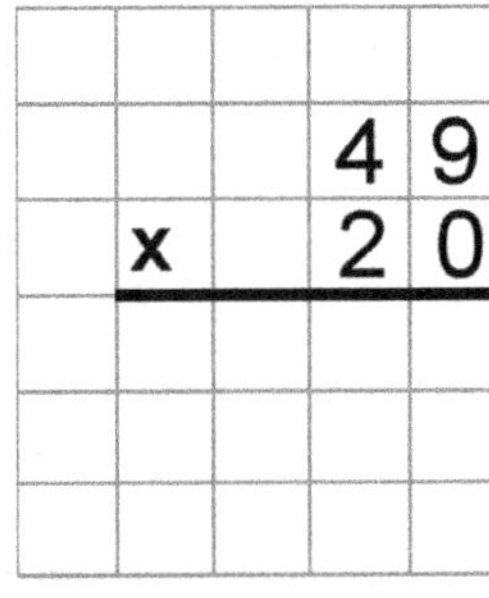

e.

38
x 13

f.

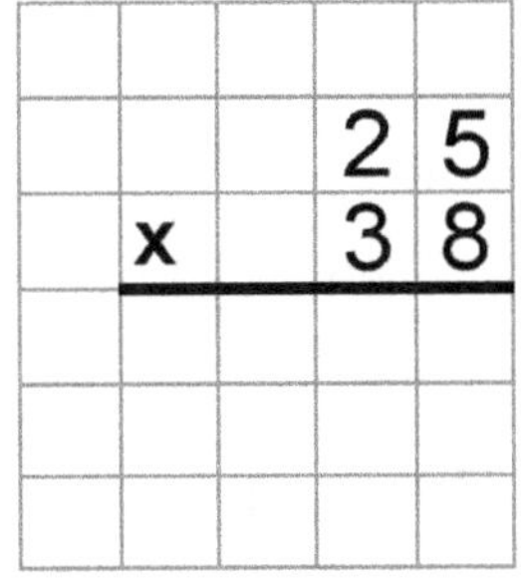

g.

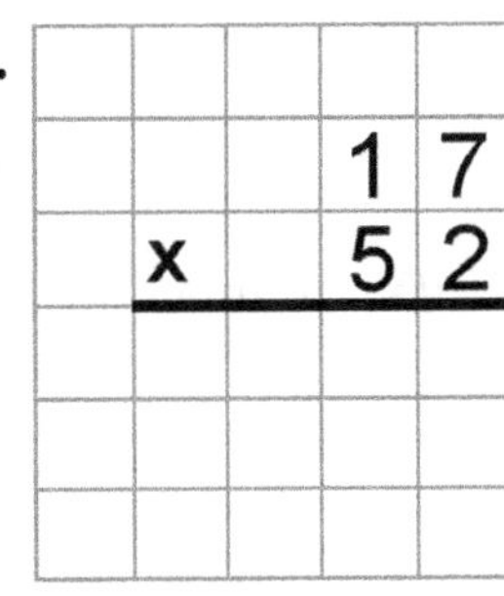

h.

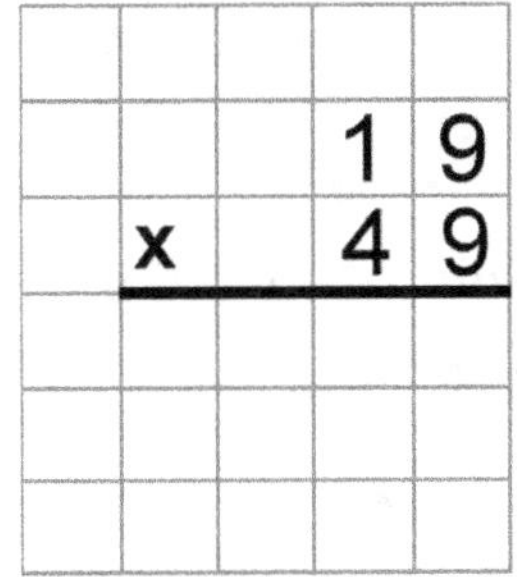

3. Multiply – but first, estimate the result. Compare your final answer to your estimated answer. If there is a big difference, you might have an error somewhere.

a. Estimate: ______ × ______

= ___________

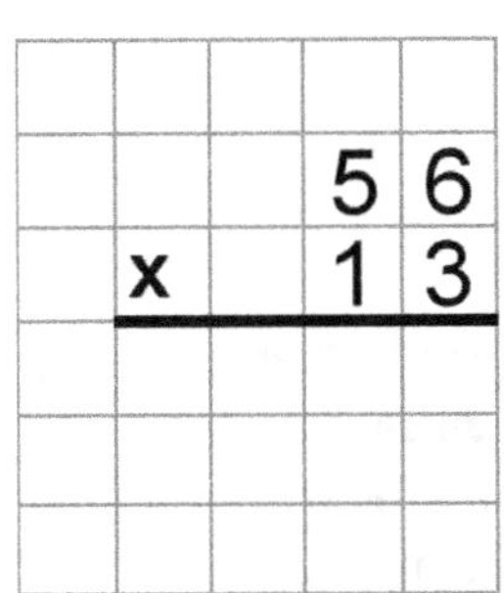

b. Estimate: ______ × ______

= ___________

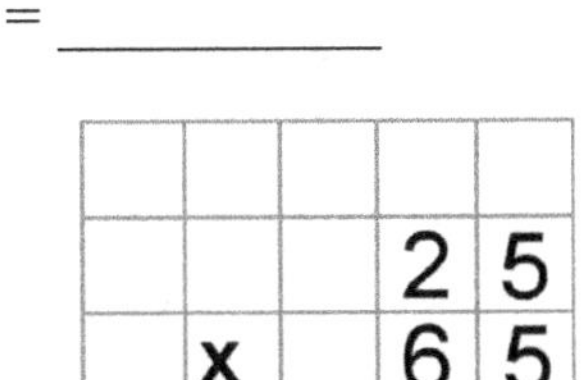

c. Estimate: ______ × ______

= ___________

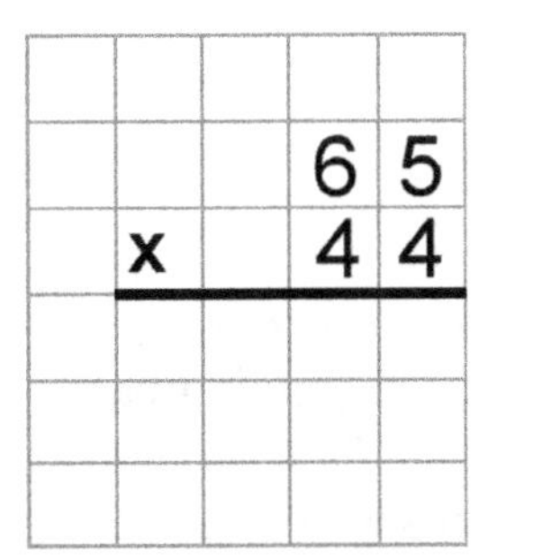

d. Estimate: ______ × ______

= ___________

44
x 14

e. Estimate: ______ × ______

= ___________

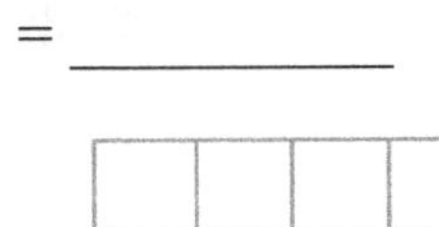

92
x 46

f. Estimate: ______ × ______

= ___________

93
x 77

4. Multiply – but first, estimate the result.

a. Estimate: ______ × ______

= __________

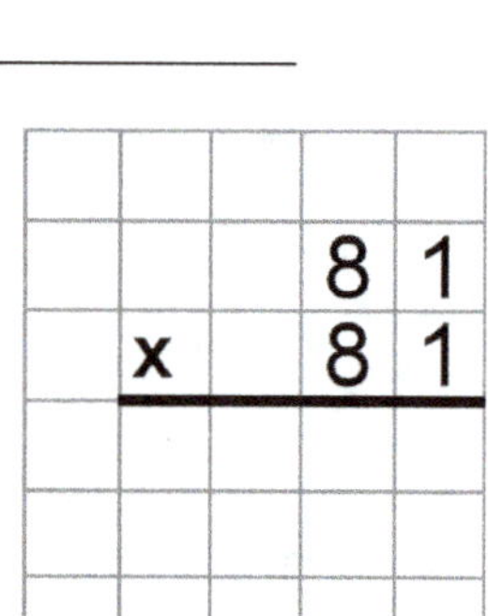

b. Estimate: ______ × ______

= __________

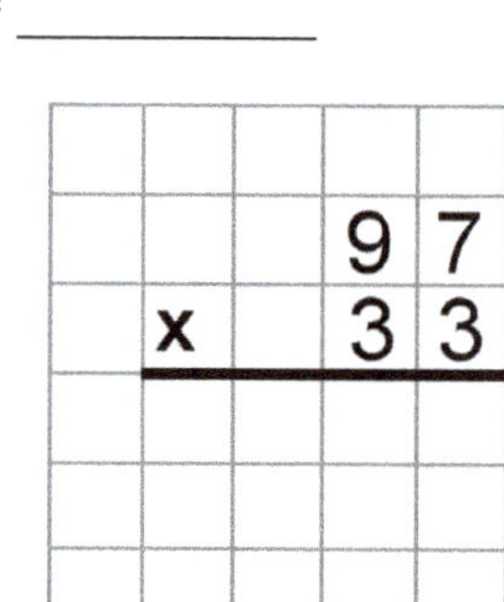

c. Estimate: ______ × ______

= __________

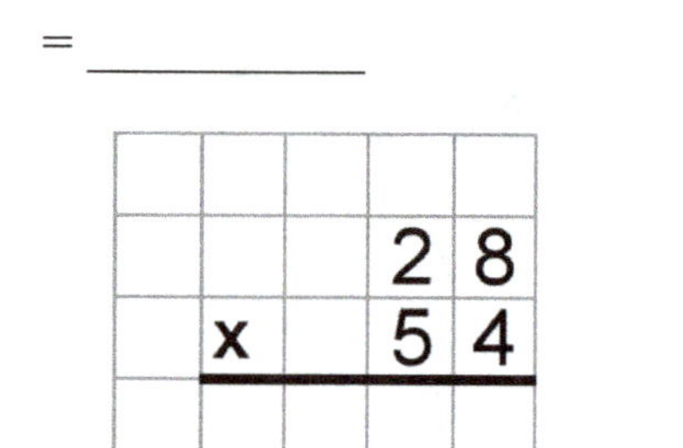

5. Solve the word problems. Write a number sentence for each one. Give more than just the final answer.

a. How many eggs is 15 dozen eggs?

____

Estimate: ________________________

b. How many minutes are there in 21 hours?

____

Estimate: ________________________

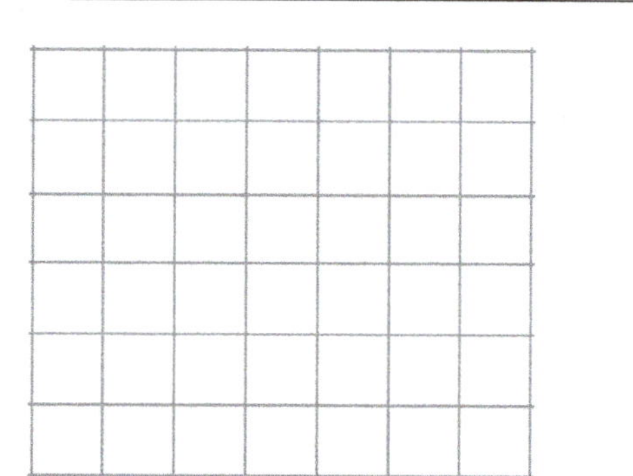

c. The 455 pupils in a school are going to a zoo by bus. One bus can seat 39 passengers. Are 11 buses enough to take them all?

____

Estimate: ________________________

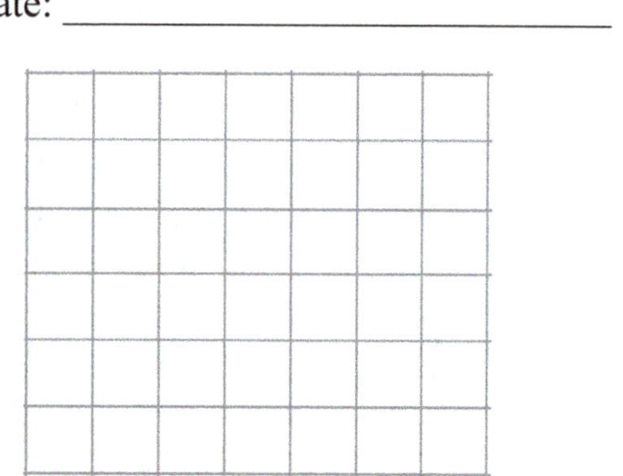

d. Each month, Brenda earns $21 for watering her neighbor's flowers. How much does she earn in a year?

____

Estimate: ________________________

6. **a.** Is the answer to 53 × 61 he same as to 51 × 63?
After all, it is just switching the ones digits.

b. What about 42 × 71 and 41 × 72?
If they are not the same, how much is the difference?

7. Find the change, if a teacher buys 15 shirts for \$17 each and pays with \$300.

8. One year has 52 weeks. Sally pays \$98 weekly in rent. How much will she pay in a year?

9. Calculate in the correct order.

a.	60 × (10 + 20) × 2 = ________
	30 × (40 − 40) × 2 = ________
b.	8 × (200 − 100) − 500 = ________
	(800 − 200) × 20 + 100 = ________

Puzzle Corner Fill in the missing numbers in these multiplications:

$$\begin{array}{r} \square\,0\,\square \\ \times\quad 3 \\ \hline 3\,1\,5 \end{array} \qquad \begin{array}{r} 6\,\square\,7 \\ \times\quad \square \\ \hline 4\,0\,0\,2 \end{array} \qquad \begin{array}{r} \square\,3\,\square \\ \times\quad 4 \\ \hline 9\,\square\,6 \end{array} \qquad \begin{array}{r} 8\,\square\,4 \\ \times\quad 5 \\ \hline 4\,\square\,7\,\square \end{array}$$

Review

1. Multiply.

a. $400 \times 3 =$ ______ $9 \times 20 =$ ______	b. $70 \times 60 =$ ______ $300 \times 11 =$ ______	c. $90 \times 900 =$ ______ $100 \times 400 =$ ______

2. Find the missing factors. Think of how many zeros you need.

a. ______ $\times 50 = 4{,}000$ ______ $\times 50 = 350$	b. $70 \times$ ______ $= 280$ $7 \times$ ______ $= 2{,}800$	c. ______ $\times 40 = 12{,}000$ ______ $\times 800 = 64{,}000$

3. Solve the equations.

a. $4 \times 30 = \underline{?} \times 3$ $\underline{?} =$ ______	b. $y \times 500 = 250 \times 4$ $y =$ ______	c. $450 + 350 = \triangle \times 20$ $\triangle =$ ______

4. Solve this problem using **estimation**.
If you earn $515 weekly, in how many weeks will you have earned more than $4,000?

5. Multiply. Estimate the answer on the line.

a. 7×48 $\approx$ ______ 	b. 6×813 $\approx$ ______ 	c. 21×18 $\approx$ ______ 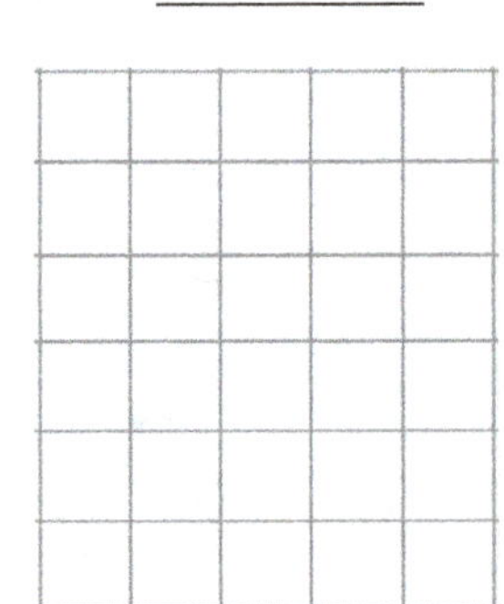	d. $4 \times 5{,}903$ $\approx$ ______

6. Fill in the table.

Roses	**1**	**2**	**3**	**4**	**5**	**6**	**7**	**8**
Price	**$0.90**							

7. Calculate in the right order.

$2 \times 98 - 8 \times 17$

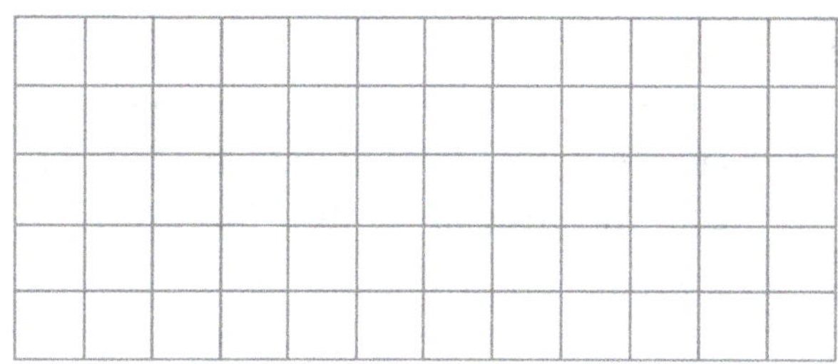

8. Solve.

a. $(1,500 - 1,000) \times 4 =$ __________

b. $(76 + 34) \times 2 \times 0 =$ __________

c. $8 \times 2 \times (3 + 2) =$ __________

d. $200 \times (500 - 400) =$ __________

9. Draw a rectangle with several parts to illustrate the multiplications. You don't have to draw accurately—a sketch is good enough.

a. 8×24

= ___ × ______ + ___ × ___

= __________

b.

```
   3 5
 × 3 9
 -----



+
------
```

10. Solve. Write a number sentence for each one, not just the answer.

a. A store owner bought 50 boxes of shirts, with 20 shirts in each box, and each shirt costs $2. What was the total cost?

____

____

b. Dad bought 8 boxes of nails for $2.35 a box. What was his change from $20?

____

____

c. Charlene bought five ice cream cones for $1.50 each. Now she has $12.50 left. How much did she have originally?

____

____

d. A huge roll of wrapping paper costs $45 but it was discounted by $8. How much do five rolls cost?

____

____

11. Solve the problems. You can use the tables to help.

a. A dog can run three miles in 15 minutes. How far could it run in 10 minutes?

b. Seven cans of tuna weigh 420 g. How much would ten cans weigh?

Math Mammoth Multiplication 2 Answer Key

Understanding Multiplication, pp. 7-9

Page 7

1. a. $2 + 2 + 2 + 2 = 4 \times 2 = 8$; $20 + 20 + 20 + 20 = 4 \times 20 = 80$
   b. $8 + 8 + 8 = 3 \times 8 = 24$; $80 + 80 + 80 = 3 \times 80 = 240$;
   c. $500 + 500 + 500 + 500 = 4 \times 500 = 2{,}000$ $120 + 120 + 120 = 3 \times 120 = 360$

2. a. $6 \times 3 = 18$; $3 \times 6 = 18$ b. $5 \times 1 = 5$; $1 \times 5 = 5$

3. a. 16, 0 b. 15, 10 c. 16, 8 d. 30, 27

4. a. 48, 0, 16 b. 300; 6,000; 12,000 c. 4,000; 1,000; 633 d. 68, 63, 200

Page 8

5. a. 6, 0 b. 200, 250 c. 3, 81

6. a. factors; product b. $4 \times 8 = 32$ c. The product is 0. d. 5, because $2 \times 6 \times \underline{5} = 60$

7. a. $3 \times 12 + 5 = 41$ eggs in all. b. Jack had: $6 \times 10 - 3 = 57$ left.
   c. $4 \times 10 + 3 \times 6 = 58$ crayons in all. d. Ernest's change was: \$50 − 3 × \$11 = \$17.
   e. $5 \times 3 + 7 \times 2 = 29$ wheels

Page 9

8.

a. $7 \times 10 = \underline{?}$ $\underline{?} = 70$	b. $\underline{?} \times 4 = 24$ $\underline{?} = 6$
c. $\underline{?} \times 2 = 18$ $\underline{?} = 9$	d. $y \times 4 = 36$ $y = 9$
e. $y \times 10 = 150$ $y = 15$	f. $y \times 12 = 60$ $y = 5$
g. $4 \times 20 = y$ $y = 80$	h. $y \times 7 = 35$ $y = 5$
i. $300 \times y = 1{,}200$ $y = 4$	j. $y \times 2 = 40$ $y = 20$

Multiplication Tables Review, pp. 10-12

Page 10

1.

1 × 5 = 5	7 × 5 = 35	1 × 10 = 10	7 × 10 = 70	1 × 11 = 11	7 × 11 = 77
2 × 5 = 10	8 × 5 = 40	2 × 10 = 20	8 × 10 = 80	2 × 11 = 22	8 × 11 = 88
3 × 5 = 15	9 × 5 = 45	3 × 10 = 30	9 × 10 = 90	3 × 11 = 33	9 × 11 = 99
4 × 5 = 20	10 × 5 = 50	4 × 10 = 40	10 × 10 = 100	4 × 11 = 44	10 × 11 = 110
5 × 5 = 25	11 × 5 = 55	5 × 10 = 50	11 × 10 = 110	5 × 11 = 55	11 × 11 = 121
6 × 5 = 30	12 × 5 = 60	6 × 10 = 60	12 × 10 = 120	6 × 11 = 66	12 × 11 = 132

Multiplication Tables Review, cont.

Page 10

2. The products (answers) 10, 20, 30, 40, 50, and 60 are found both in the table of 5 and in the table of 10. That is because 10 = 2 × 5, so any product of 10 also is a product of 5.

1 × 2 = 2	7 × 2 = 14	1 × 4 = 4	7 × 4 = 28	1 × 8 = 8	7 × 8 = 56
2 × 2 = 4	8 × 2 = 16	2 × 4 = 8	8 × 4 = 32	2 × 8 = 16	8 × 8 = 64
3 × 2 = 6	9 × 2 = 18	3 × 4 = 12	9 × 4 = 36	3 × 8 = 24	9 × 8 = 72
4 × 2 = 8	10 × 2 = 20	4 × 4 = 16	10 × 4 = 40	4 × 8 = 32	10 × 8 = 80
5 × 2 = 10	11 × 2 = 22	5 × 4 = 20	11 × 4 = 44	5 × 8 = 40	11 × 8 = 88
6 × 2 = 12	12 × 2 = 24	6 × 4 = 24	12 × 4 = 48	6 × 8 = 48	12 × 8 = 96

The products (answers) 8, 16, and 24 are in the tables of 2, 4, and 8. Because 8 = 2 × 4, any product of 8 also is a product of 2 and of 4. For example, 24 = 3 × 8 = 6 × 4 = 12 × 2.

Page 11

3.

1 × 3 = 3	7 × 3 = 21	1 × 6 = 6	7 × 6 = 42	1 × 9 = 09	7 × 9 = 63
2 × 3 = 6	8 × 3 = 24	2 × 6 = 12	8 × 6 = 48	2 × 9 = 18	8 × 9 = 72
3 × 3 = 9	9 × 3 = 27	3 × 6 = 18	9 × 6 = 54	3 × 9 = 27	9 × 9 = 81
4 × 3 = 12	10 × 3 = 30	4 × 6 = 24	10 × 6 = 60	4 × 9 = 36	10 × 9 = 90
5 × 3 = 15	11 × 3 = 33	5 × 6 = 30	11 × 6 = 66	5 × 9 = 45	11 × 9 = 99
6 × 3 = 18	12 × 3 = 36	6 × 6 = 36	12 × 6 = 72	6 × 9 = 54	12 × 9 = 108

The products (answers) 6, 12, 18, 24, 30, and 36 are in tables of 3 and 6. Because 6 = 2 × 3, any product of 6 is also a product of 3. For example, 30 = 5 × 6 = 10 × 3.

In the table of nine, the tens digits of the products (colored red) go from 0 to 9. Then there is a 9 (of 99) and then a 10 (of 108). From then on they would once again continue in order. The ones digits start at 9 and decrease one by one to 0, then start over again.

4.

1 × 7 = 7	7 × 7 = 49	1 × 12 = 12	7 × 12 = 84
2 × 7 = 14	8 × 7 = 56	2 × 12 = 24	8 × 12 = 96
3 × 7 = 21	9 × 7 = 63	3 × 12 = 36	9 × 12 = 108
4 × 7 = 28	10 × 7 = 70	4 × 12 = 48	10 × 12 = 120
5 × 7 = 35	11 × 7 = 77	5 × 12 = 60	11 × 12 = 132
6 × 7 = 42	12 × 7 = 84	6 × 12 = 72	12 × 12 = 144

5. a. 7, 4, 8 b. 8, 5, 9 c. 8, 6, 7 d. 8, 6, 7 e. 9, 7, 8 f. 9, 6, 7
g. 7, 9, 8 h. 9, 8, 6 i. 5, 9, 3 j. 12, 5, 6 k. 6, 12, 7 l. 9, 2, 4

Multiplication Tables Review, cont.

Page 12

6.

×	0	1	2	3	4	5	6	7	8	9	10	11	12
0	0	0	0	0	0	0	0	0	0	0	0	0	0
1	0	1	2	3	4	5	6	7	8	9	10	11	12
2	0	2	4	6	8	10	12	14	16	18	20	22	24
3	0	3	6	9	12	15	18	21	24	27	30	33	36
4	0	4	8	12	16	20	24	28	32	36	40	44	48
5	0	5	10	15	20	25	30	35	40	45	50	55	60
6	0	6	12	18	24	30	36	42	48	54	60	66	72
7	0	7	14	21	28	35	42	49	56	63	70	77	84
8	0	8	16	24	32	40	48	56	64	72	80	88	96
9	0	9	18	27	36	45	54	63	72	81	90	99	108
10	0	10	20	30	40	50	60	70	80	90	100	110	120
11	0	11	22	33	44	55	66	77	88	99	110	121	132
12	0	12	24	36	48	60	72	84	96	108	120	132	144

7. $90 - 8 \times 7 = 34$. So, 34 students went by bus.

8. a. 33 b. 26 c. 8 d. 70 e. 63 f. 26

9.

	a. $2 \times 6 = 4 \times \underline{3}$	b. $6 \times 6 = 4 \times \underline{9}$
c. $3 \times 10 = 6 \times \underline{5}$	d. $2 \times 20 = 10 \times \underline{4}$	e. $5 \times 12 = 6 \times \underline{10}$

Scales Puzzles, pp. 13-16

Page 13

1. a. 7 b. 6 c. 6 d. 2 e. 7 f. 4 g. 2 h. 7

Page 14

2. a. 11 b. 8 c. 13 d. 5

3. a. Hint: from the first balance we learn that one circle + one rectangle weigh 14. In the second balance, we have two circles and one rectangle, and of those, one rectangle and one circle together still weigh 14. This means that the one circle must weigh 6.
Solution: 1 rectangle weighs 8 and 1 circle weighs 6.

b. Hint: guess and check using small numbers. Look at the second scales, and in it, try for example one circle being 2. Then you will notice that that will not work! The circle has to be a bigger number. Check to see if the circle is 4. Then the diamond would be 0. That will not work either. Check to see if the circle is 6. Then the diamond would be 4. Now go to the first scale and check, if the circle = 6 and the diamond = 4 works there. If not, change your guess.
Solution: 1 circle weighs 5 and 1 diamond weighs 2.

Page 15

4. a. 1 circle weighs 4 and 1 square weighs 17. b. 1 square weighs 5 and 1 triangle weighs 3.
c. 1 square weighs 3 and 1 circle weighs 11. d. 1 circle weighs 2 and 1 triangle weighs 3.

5. a. 70 b. 29 c. 60 d. 151 e. 9 f. 21 g. 18 h. 17 i. 8

Multiplying by Whole Tens and Hundreds, pp. 17-20

Page 17

1. a. 3,150; 35,600; 3,500   b. 620,000; 12,000; 13,000   c. 250,000; 38,000; 50,000

2. a. 160; 80; 100   b. 1400; 1,000; 2,200   c. 240; 700; 1,800   d. 320; 8,400; 1,080

Page 18

3.

a. 20 × 7	b. 20 × 5	c. 200 × 8	d. 200 × 25
= 10 × 2 × 7	= 10 × 2 × 5	= 100 × 2 × 8	= 100 × 2 × 25
= 10 × 14	= 10 × 10	= 100 × 16	= 100 × 50
= 140	= 100	= 1600	= 5000

4. A = 20 ft × 15 ft = 300 ft^2

5. A = 15 ft × 200 ft = 3,000 ft^2

6. One truckload costs 5 × \$20 + \$30 = \$130. Four truckloads cost 4 × \$130 = \$520.

Page 19

7. a. 120; 160   b. 420; 550   c. 720; 450   d. 660; 480   e. 1,800; 2,800
f. 4,200; 6,600   g. 2,400; 4,500   h. 3,300; 7,200   i. 1,320; 2400

8. a. 1,800; 21,000   b. 4,800; 27,000   c. 20,000; 40,000   d. 64,000; 100,000   e. 10,000; 1,200   f. 240,000; 99,000

Page 20

9. One hour has 60 minutes.
How many minutes are in 12 hours? 12 × 60 min = 720 minutes
How many minutes are in 24 hours? 2 × 720 min = 1,440 minutes (double the previous result)

10. One hour has 60 minutes, and one minute has 60 seconds.
How many seconds are there in one hour? 60 × 60 sec = 3,600 seconds

11. a. 8 × \$30 = \$240   b. 40 × \$30 = \$1,200
c. Guess and check: 2 × \$240 = \$480; 4 × \$240 = \$960; 5 × \$240 = \$1,200; so five days.

12. a. 120; 9   b. 8; 120   c. 10; 90   d. 160; 9   e. 50; 700   f. 70; 600

Puzzle Corner: 600 × 50 = 6 × 100 × 5 × 10 = 100 × 10 × (6 × 5) = 1000 × 30 = 30,000.

Multiply in Parts 1, pp. 21-23

Page 21

1.

a. 6 × 27 (20 + 7)	b. 5 × 83 (80 + 3)	c. 9 × 34 (30 + 4)
6 × 20 and 6 × 7	5 × 80 and 5 × 3	9 × 30 and 9 × 4
120 and 42	400 and 15	270 and 36
= 162	= 415	= 306
d. 3 × 99	e. 7 × 65	f. 4 × 58
3 × 90 and 3 × 9	7 × 60 and 7 × 5	4 × 50 and 4 × 8
270 and 27	420 and 35	200 and 32
= 297	= 455	= 232

Multiply in Parts 1, cont.

Page 22

2.

a. $9 \times 17 = 9 \times 10 + 9 \times 7$

$= 90 + 63 = 153$

b. $6 \times 29 = 6 \times 20 + 6 \times 9$

$= 120 + 54 = 174$

c. $7 \times 33 = 7 \times 30 + 7 \times 3$

$= 210 + 21 = 231$

3.

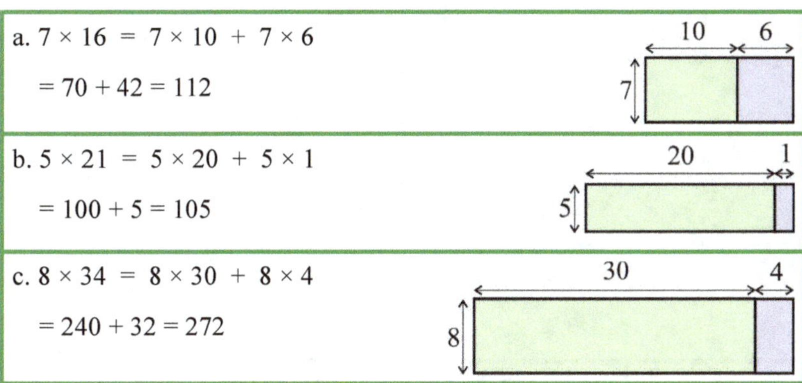

Page 23

4.

a. 6 × 19	b. 3 × 73	c. 4 × 67
6 × 10 → 60 6 × 9 → + 54 114	3 × 70 → 210 3 × 3 → + 9 219	4 × 60 → 240 4 × 7 → + 28 268
d. 5 × 92	**e. 9 × 33**	**f. 7 × 47**
5 × 90 → 450 5 × 2 → + 10 460	9 × 30 → 270 9 × 3 → + 27 297	7 × 40 → 280 7 × 7 → + 49 329

5. a. 65 b. 135 c. 165 d. 168 e. 88 f. 357

6. a. > b. > c. >

7. a. Jack's total cost was 8 × \$14 = 8 × \$10 + 8 × \$4 = \$80 + \$32 = <u>\$112</u>.
   b. There were 9 × 14 + 56 = <u>182</u> seats in all.
   c. The cost of the hammer is 3 × \$17 = <u>\$51</u>.

Multiply in Parts 2, pp. 24-25

Page 24

1.

a. 3 × 127 (100 + 20 + 7) 3 × 100 and 3 × 20 and 3 × 7 300 and 60 and 21 = 381	b. 5 × 243 (200 + 40 + 3) 5 × 200 and 5 × 40 and 5 × 3 1,000 and 200 and 15 = 1,215
d. 4 × 6,507 (6000 + 500 + 7) 4 × 6,000 and 4 × 500 and 4 × 7 24,000 and 2,000 and 28 = 26,028	
e. 5 × 4,813 5 × 4,000 + 5 × 800 + 5 × 10 + 5 × 3 20,000 + 4,000 + 50 + 15 = 24,065	

Page 25

2.

a. 4 × 128	b. 8 × 151	c. 3 × 452
4 × 100 → 400 4 × 20 → 80 4 × 8 → + 32 512	800 400 + 8 1208	1200 150 + 6 1356
d. 6 × 3,217	e. 8 × 2,552	f. 6 × 1,098
18000 1200 600 + 42 19302	16000 4000 400 + 16 20416	6000 0 540 + 48 6588

3. a. In half a year, Dad pays 6 × $138 = $600 + $180 + $48 = $828.

b. The perimeter is 4 × 255 cm = 800 cm + 200 cm + 20 cm = 1,020 cm.

c. The bigger roll contains 5 × 56 cm = 250 cm + 30 cm = 280 cm of material.
In total the rolls have 56 cm + 280 cm = 336 cm of material.

Multiply in Parts – Area Model, pp. 26-27

Page 26

1.

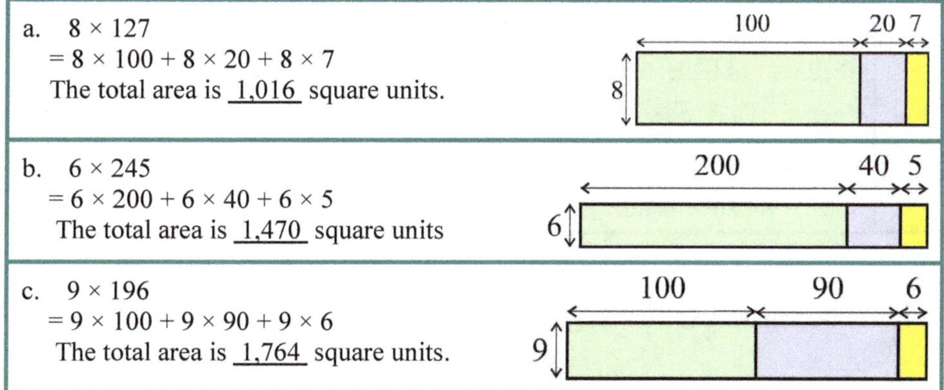

2. a. 6 × 6 = 9 × 4   b. 12 × 10 = 5 × 24
   c. 20 + 20 = 4 × 10   d. 6000 = 30 × 200
   e. 120 − 75 = 5 × 9   f. 750 + 750 = 5 × 300

Page 27

3.

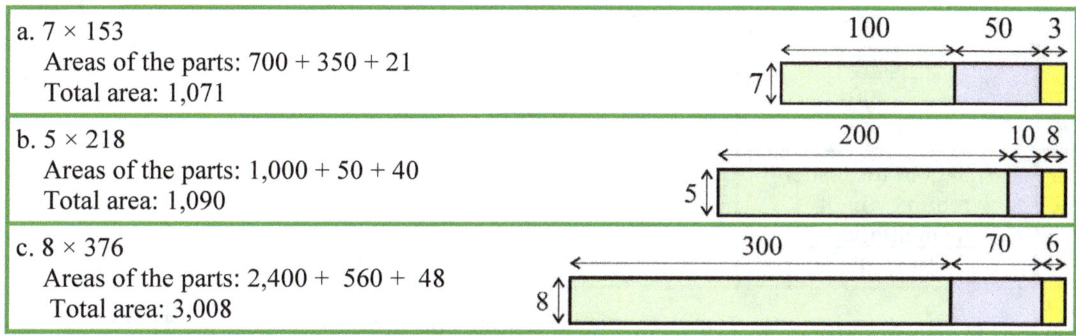

4. a. Susie orders 5 × 72 = 360 flowers in 5 weeks.
   b. It costs her 5 × $70 = $350 for five weeks of orders.

Multiplying Money Amounts, pp. 28-29

Page 28

1.

a. 6 × $11.85	b. 5 × $2.93
6 × $11 → $66.00 6 × $0.80 → 4.80 6 × $0.05 → + 0.30 $71.10	5 × $2 → $10.00 5 × $0.90 → 4.50 5 × $0.03 → + 0.15 $14.65
c. 7 × $3.75	**d. 8 × $10.95**
7 × $3 → $21.00 7 × $0.70 → 4.90 7 × $0.05 → + 0.35 $26.25	8 × $10 → $80.00 8 × $0.90 → 7.20 8 × $0.05 → + 0.40 $87.60

Multiplying Money Amounts, cont.

Page 28

2.

a. 6 × \$2.80 \$12 + \$4.80 = \$16.80 (6 × \$2) (6 × \$0.80)	b. 5 × \$4.70 \$20 + \$3.50 = \$23.50 (5 × \$4) (5 × \$0.70)
c. 4 × \$12.50 \$48 + \$2 = \$50	d. 7 × \$5.61 \$35 + \$4.20 + \$0.07 = \$39.27

Page 29

3. a. Her total cost was 5 × \$2.00 + 5 × \$0.70 = \$10.00 + \$3.50 = \$13.50.
   b. Her change was \$20.00 − \$13.50 = \$6.50.

4. a. The cost for four is 4 × 23.50 = 4 × \$20 + 4 × \$3 + 4 × \$0.50 = \$80 + \$12 + \$2 = \$94.
   b. Their change is \$100.00 − \$94.00 = \$6.00.

5.

a. Start at 80. Add 40 each time:	b. Start at 42,000. Subtract 3,000 each time:	c. Start at 1. Add 5 each time:
120 160 200 240 280 320 360	42,000 39,000 36,000 33,000 30,000 27,000 24,000	1 6 11 16 21 26 31 36 41
What does this pattern remind you of? The multiplication table of 4.	What does this pattern remind you of? The multiplication table of 3.	

6. When you add 5 to a number, it changes parity (goes from even to odd, or from odd to even). In other words, if you add 5 to an odd number, you get an even number, and vice versa. In fact, the same happens when you repeatedly add any odd number.

Estimating in Multiplication, pp. 30-31

Page 30

1. Answers may vary. Estimating is not an "exact science".
   a. 5 × 70 = 350 b. 11 × 60 = 660 c. 120 × 8 = 960 d. 30 × 50 = 1,500 e. 7 × \$4 = \$28
   f. 8 × \$12 = \$96 g. 25 × \$40 = \$1,000 h. 9 × 20 = 180 or 10 × 17 = 170 i. 60 × 900 = 54,000

2. a. 20 × \$45 = \$900; however since this is about shopping, it might be better not to round down so much, and estimate the cost as 24 × \$45, which you can calculate in two parts: 20 × \$45 and 4 × \$45, which is \$900 + \$180 = \$1,080.
   b. 500 × 20¢ = 10,000¢ = \$100
   c. 200 × \$1.50 = \$200 + \$100 = \$300
   d. Tennis balls: 6 × \$3 = \$18; Rackets: 2 × \$12 = \$24; Total cost: \$42.

Page 31

3. a. Bill can buy 5 ads. Round to \$350. Then add: two ads is \$700, four ads is \$1,400, six ads is \$2,100. So he can not afford six ads. Five ads would be \$1,400 + \$350 = \$1,750.
   b. Round the rate to \$3 per hour. Since 8 × \$3 = \$24, she can rent them for about 8 hours.
   c. Beans: 8 × \$0.30 = \$2.40; Lentils: 5 × \$0.40 = \$2. So it is cheaper to buy 5 bags of lentils.
   d. Round the cost of string to \$0.20 per foot. Round the number of children to 30.
   Cost per one child is about 8 × \$0.20 = \$1.60. The total cost is about 30 × \$1.60 = \$48.

Multiply in Columns—the Easy Way, pp. 32-34

Page 32

1. a. 456 b. 445 c. 301 d. 312 e. 415 f. 564 g. 288 h. 287

Page 33

2. a. 822 b. 872 c. 2,191 d. 2,256 e. 3,608 f. 2,085
g. 6,240 h. 1,944 i. 5,562 j. 1,698 k. 2,056 l. 3,040

3. a. 6 b. 90 c. 90

Page 34

4. a. 581 b. 565

5. a. (236 − \$40) × 7 = \$196 × 7 = \$1,372
b. 992 ft

6. a. 40 b. 40 c. 700
d. 12 e. 100 f. 80

Puzzle corner. a. 172; 1204 b. 358; 32 c. 709; 00

Multiply in Columns—the Easy Way, part 2, pp. 35-37

Page 35

1. a. 10,628 b. 30,528 c. 24,318 d. 56,712 e. 11,212 f. 26,460

Page 36

2. a. 9,930 b. 18,495 c. 22,960 d. 36,702 e. 12,445 f. 34,408

3. a. The total distance is 2 × 2 × 3,820 = 4 × 3,820 = 15,280 feet.
b. 55 + 4 × 55 = 275 marbles in total.

Page 37

4. a. \$5.49 b. \$35.48 c. \$80.50 d. \$61.50
e. \$60.30 f. \$58.94 g. \$82.60 h. \$318.48

5. a. Estimate: 10 × 1.57 = 15.70 or 10 × 1.60 = 16.00. Total cost: 9 × \$1.57 = \$14.13
b. Estimate: 8 × 2.3 = 18.40. Change: \$20 − 8 × \$2.28 = \$20 − \$18.24 = \$1.76

Multiply in Columns, the Standard Way, pp. 38-41

Page 38

1.

a.

	2	
	5	3
x		8
4	2	4

	5	3
x		8
	2	4
4	0	0
4	2	4

b.

	2	
	8	8
x		3
2	6	4

	8	8
x		3
	2	4
2	4	0
2	6	4

Page 39

2.

a.

	2	
	7	9
x		3
2	3	7

	7	9
x		3
	2	7
2	1	0
2	3	7

b.

	4	
	1	8
x		5
	9	0

	1	8
x		5
	4	0
	5	0
	9	0

Multiply in Columns, the Standard Way, cont.

Page 39

3. a. 306 b. 57 c. 124 d. 322 e. 396 f. 351
   g. 261 h. 134 i. 180 j. 432 k. 204 l. 92

4. a. Three chairs: $3 \times \$48 = \$144$. Six chairs: $2 \times \$144 = \$288$ (double the previous result).
   b. In five days, you earn $5 \times \$77 = \$385$.
   In ten days, you earn $10 \times \$77 = \$770$.

Page 40

5.

a.
```
  1 2 3        1 2 3
x     8      x     8
  9 8 4          2 4
               1 6 0
               8 0 0
               9 8 4
```

b.
```
  2 2
  2 7 9        2 7 9
x     3      x     3
  8 3 7          2 7
               2 1 0
               6 0 0
               8 3 7
```

c.
```
   4 6 3         4 6 3
x      5     x       5
 2 3 1 5           1 5
                 3 0 0
               2 0 0 0
               2 3 1 5
```

d.
```
  1 5 6        1 5 6
x     6      x     6
  9 3 6          3 6
               3 0 0
               6 0 0
               9 3 6
```

Page 41

6. a. 924 b. 3,542 c. 1,390 d. 2,233 e. 864 f. 7,281
   g. 861 h. 734 i. 10,872 j. 7,542 k. 24,708 l. 47,970

7. Perimeter: 2×9 ft $+ 2 \times 28$ ft $= 18$ ft $+ 56$ ft $= 74$ ft.
   Area: 9 ft $\times$ 28 ft $= 252$ sq. ft.

8. a. Every time she multiplies she carries the ones digit instead of the tens digit.
   The real answers are: 234, 342, 532.
   b. Andy does not carry. He writes the tens digit that he should carry as part of the answer.
   The real answers are: 84, 225, 645.

Multiplying in Columns, Practice, pp. 42-43

Page 42

1. a. 387
   b. Estimation: $8 \times 70 = 560$; Exact: 576
   c. Estimation: $3 \times 770 = 2{,}310$; Exact: 2,313
   d. Estimation: $5 \times 800 = 4{,}000$; Exact: 4,095
   e. Estimation: $4 \times 2{,}500 = 10{,}000$; Exact: 10,084
   f. Estimation: $3 \times 9{,}000 = 27{,}000$; Exact: 26,136

2. It has $3 \times 187 = 561$ pages.

Page 43

3. Five packages have $5 \times 250 = 1{,}250$ sheets, which is not enough. $6 \times 250 = 1{,}500$, which is enough.
   So, she needs to buy six packages.

4. a. Four buses could seat $4 \times 43 = 172$ students.
   b. Seven buses could seat $7 \times 43 = 301$ students.
   c. They need eight buses.

5. a. 80 b. 3 c. 30

Multiplying in Columns, Practice, cont.

Page 43

6. a. Estimation: $5 \times 200 = 1{,}000$; Exact: 980
   b. Estimation: $9 \times 200 = 1{,}800$; Exact: 1,845
   c. Estimation: $9 \times 10{,}000 = 90{,}000$; Exact: 88,263
   d. Estimation: $6 \times 5{,}000 = 30{,}000$; Exact: 28,860

Puzzle Corner: $117 \times 4 = 468$; $174 \times 5 = 870$; $138 \times 7 = 966$; $3{,}219 \times 3 = 9{,}657$

Order of Operations Again, pp. 44-46

Page 44

1. a. First calculate the sum <u>23</u> + <u>31</u>.
   b. Next multiply that sum by <u>9</u>.
   c. Subtract that result from <u>650</u>.
   d. Lastly add <u>211</u> to the subtraction result.
   Result: 375

2. a. 1,800; 150 b. 210; 2,000 c. 3,200; 0 d. 2,800; 5,600

3. a. 100; 900 b. 3,500; 700 c. 600; 1,400

4. a. 100 b. 900
   c. 5,800 d. 230
   e. 860 f. 2,700

Page 45

5. a. 316 b. 2,155 c. 1,124

6. a. $4 \times \$2 + 3 \times \$3 = \$17$
   b. $4 \times (\$2 + \$3) = \$20$
   c. $\$50 - 5 \times \$3 - 5 \times \$2 = \$25$

Page 46

7. a. N = 3 b. N = 12 c. N = 3
   d. N = 1,500 e. N = 9 f. N = 4

8. a. Her change was $\$200 - 7 \times \$25$ = <u>$25</u>.
   b. The total weight is 8×3 kg + 15×2 kg = <u>54 kg</u>.
   c. The second building is $3 \times 9 \times 9$ ft = <u>243 feet tall</u>.
   d. $2 \times \$18 - 9 \times \$3 = \$9$. The nine cans of cat food is $9 cheaper than two boxes of cat food.

Puzzle Corner: a. $7 \times (2 + 8) = 70$ b. $80 - 5 \times (10 - 5) = 55$ c. $(4 + 8) \times 5 - 20 = 40$

Money and Change, pp. 47-49

Page 47

1. a. Estimation: $4 \times \$5 = \$20$; answer $18.20
   b. Estimation: $4 \times \$10 = \$40$; answer $38.80
   c. Estimation: $7 \times \$5 = \$35$; answer $34.37
   d. Estimation: $6 \times \$1 = \$6$; answer $4.92
   e. Estimation: $7 \times \$13 = \$91$; answer $87.71
   f. Estimation: $5 \times \$43 = \$215$; answer $215.75

2. Estimates vary. For example: $8 \times \$1.40 = \$11.20$. Change: $8.80.
   Solution: $8.88. Multiply in columns or in parts $8 \times \$1.39 = \$11.12$. Then, subtract $\$20 - \$11.12 = \$8.88$.

Page 48

3.

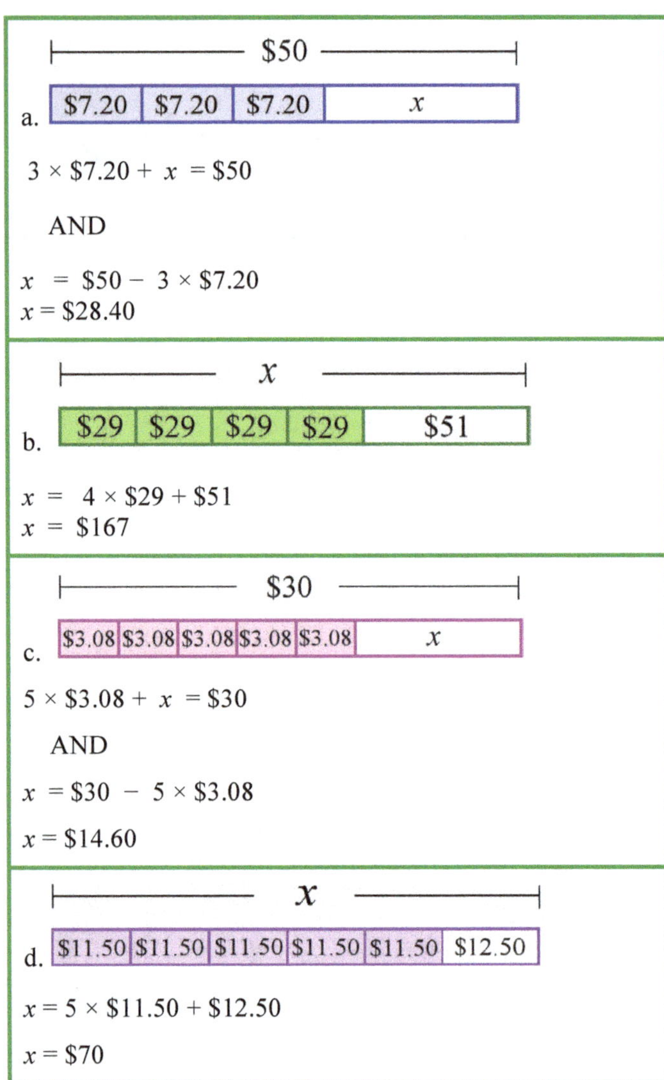

Page 49

4. Notice that you need five packs because five packs will contain 20 bottles.
Estimation: 5 × $3 = $15.
Calculation: 5 × $2.76 = $13.80.

5. Estimation: 20 × $0.15 = 300 cents = $3; 10 × $1 = $10; Total $13.
Calculation: 20 × $0.15 = 300 cents = $3. The student can solve 10 × $1.09 by multiplying in parts: 10 × $1 is $10, and 10 × 9 cents = 90 cents. The notebooks cost $10.90, and the total cost is $13.90.

6. Estimation: 4 × $10 + $65 + $26 = $40 + $65 + $26 = $131.
Calculations: 4 × $9.80 = $39.20. $39.20 + $65 + $25.80 = $130

Puzzle corner. $20. The price of one wheelbarrow was $125 because 8 × $125 = $1000.

So Many of the Same Thing, pp. 50-52

Page 50

1. a.

Miles	45	90	135	180	225	270	315	360	405	450
Hours	1	2	3	4	5	6	7	8	9	10

b.

Dollars	$5.10	$10.20	$15.30	$20.40	$25.50	$30.60	$35.70	$40.80	$45.90	$51.00
Meters	1	2	3	4	5	6	7	8	9	10

c.

Dollars	$1.50	$3.00	$4.50	$6.00	$7.50	$9.00	$10.50	$12.00	$13.50	$15.00
Cans	1	2	3	4	5	6	7	8	9	10

d.

Dollars	$30	$60	$90	$120	$150	$180	$210	$240	$270	$300
Buckets	2	4	6	8	10	12	14	16	18	20

e.

Feet	40	80	120	160	200	240	280	320	360	400
Minutes	10	20	30	40	50	60	70	80	90	100

f.

Tires	Minutes
1	15
2	30
3	45
4	60
5	75
6	90

g.

Days	Scarves
3	1
6	2
9	3
12	4
15	5
18	6

h.

Hours	Dollars
1	$15
2	$30
3	$45
4	$60
5	$75
6	$90

Two identical bags contain 12 kg of potatoes in total. How many bags would you need to have 30 kg of potatoes?

To solve this problem, you can make a chart as the one on the right:

1 bag	6 kg
2 bags	12 kg
3 bags	3 × 6 kg = 18 kg
5 bags	5 × 6 kg = 30 kg

The total number of chocolates in five identical boxes was 30. How many chocolates would two boxes contain?

First find out how many in ONE box:

1 box	6 chocolates
2 boxes	12 chocolates
5 boxes	30 chocolates

So Many of the Same Thing, cont.

Page 51

2. a.

1 flower	$3
5 flowers	$15
6 flowers	$18

b.

1 can	200 g
3 cans	600 g
4 cans	800 g

c.

1 lure	$2
3 lures	$6
7 lures	$14

d.

1 episode	30 min
3 episodes	90 min
5 episodes	150 min

e.

1 sit-up	2 sec
5 sit-ups	10 sec
30 sit-ups	60 sec

Page 52

2. f.

1 notebook	$2
7 notebooks	$14
10 notebooks	$20

g.

1 day	30 pages
4 days	120 pages
10 days	300 pages

h.

1 pair	$0.75
6 pairs	$4.50
30 pairs	$22.50

3. a.

5 cars	$35.50
1 car	$7.10
4 cars	$28.40

b.

5 rows	100 minutes
1 row	20 minutes
9 rows	180 minutes = 3 hours

c. 25 minutes.
Solution: Elaine can run 4 times around a track in an hour. This means she takes 15 minutes to run around the track. Today she ran three times. She took 45 minutes for that. Then walked the fourth time. All that took 10 minutes longer than on her normal days. This means she took 1 hour 10 minutes. So, if running took 45 minutes, and in total she used 1 h 10 min, then the walking time is the difference of those, which is 25 minutes.

Multiplying Two-Digit Numbers in Parts, pp. 53-57

Page 53

1. a. $23 \times 31 = 20 \times 30 + 20 \times 1 + 3 \times 30 + 3 \times 1 = 600 + 20 + 90 + 3 = 713$ square units.
   b. $28 \times 45 = 20 \times 40 + 20 \times 5 + 8 \times 40 + 8 \times 5 = 800 + 100 + 320 + 40 = 1{,}260$ square units.
   c. $35 \times 27 = 30 \times 20 + 30 \times 7 + 5 \times 20 + 5 \times 7 = 600 + 210 + 100 + 35 = 945$ square units.

Page 54

2.

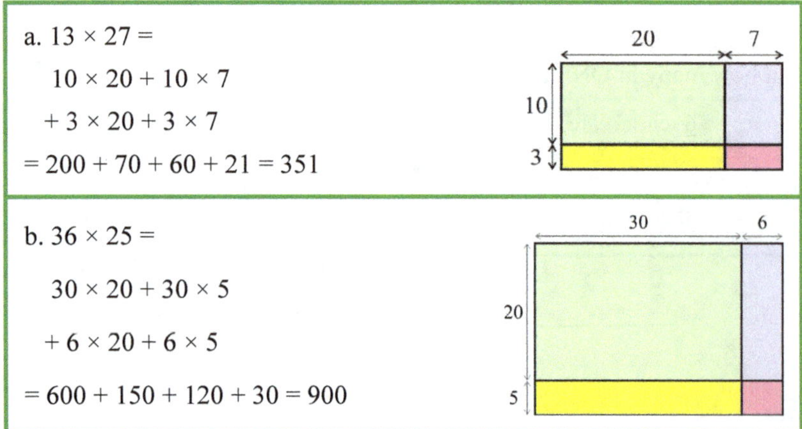

Page 54

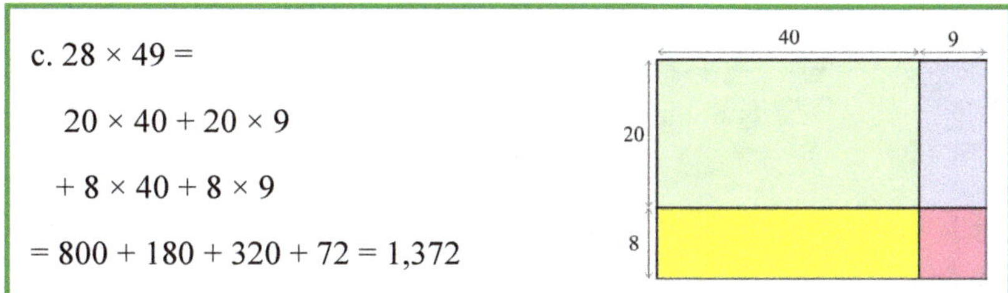

c. 28 × 49 =

20 × 40 + 20 × 9

+ 8 × 40 + 8 × 9

= 800 + 180 + 320 + 72 = 1,372

Page 55

3.

a.	87 × 15
5 × 7 →	35
5 × 80 →	400
10 × 7 →	70
10 × 80 →	+ 800
	1305

b.	24 × 71
1 × 4 →	4
1 × 20 →	20
70 × 4 →	280
70 × 20 →	+ 1400
	1704

c.	38 × 92
2 × 8 →	16
2 × 30 →	60
90 × 8 →	720
90 × 30 →	+ 2700
	3496

d.	52 × 65
5 × 2 →	10
5 × 50 →	250
60 × 2 →	120
60 × 50 →	+ 3000
	3380

Page 56

4.

a.	b.	c.	d.
55 × 12	81 × 64	73 × 80	99 × 11
10	4	0	9
100	320	0	90
50	60	240	90
+ 500	+ 4800	+ 5600	+ 900
660	5184	5840	1089

5.

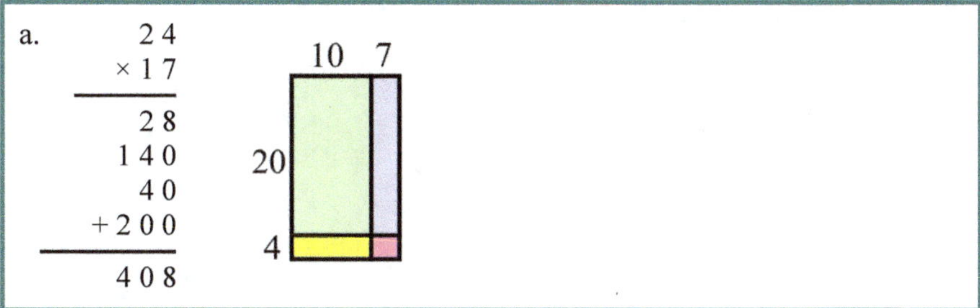

a.
24
× 17
28
140
40
+ 200
408

Page 56

5.

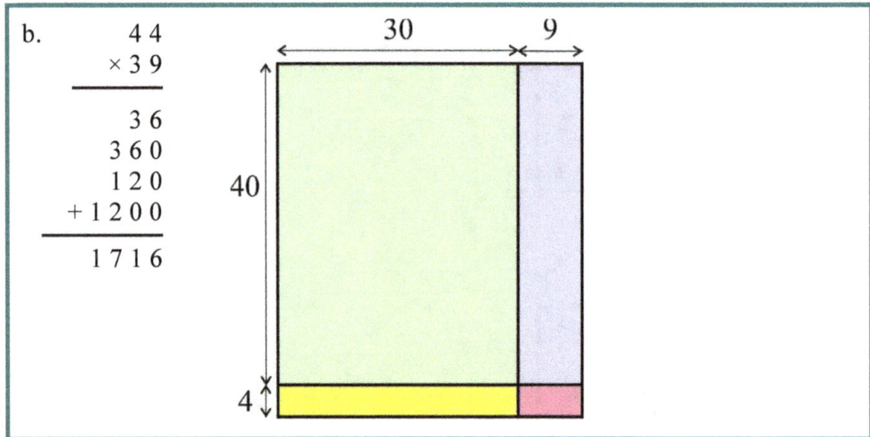

Page 57

6.

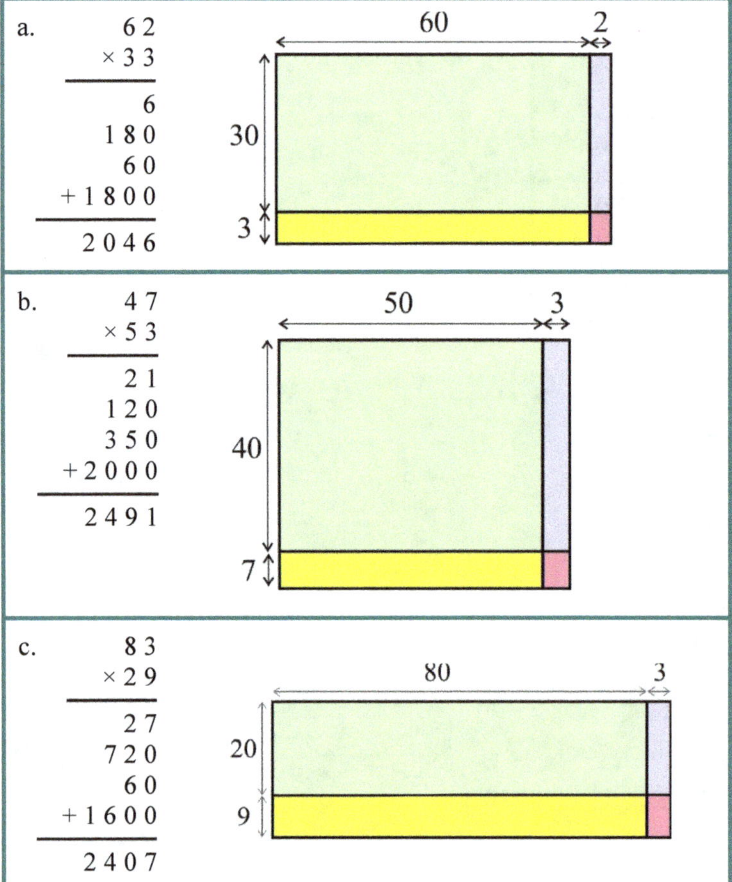

Multiply by Whole Tens in Columns, pp. 58-59

Page 58

1. a. 5,220
   b. 2,040 (first multiply 51 × 4)
   c. 1,980 (First multiply 66 × 3)

2. a. 20 × 65 kg = 1,300 kg
   b. 4 × 25 = 100 apples in each crate
   c. He got three crates, which is 300 apples because a crate weighs 20 kg and there are 100 apples per crate.

Page 59

3. a. 36,800 b. 25,500 c. 4,360
   d. 8,400 e. 134,000 f. 64,800

4. In his five-day work week he drives 5 × 250 km = <u>1,250 km</u>. In a month he drives 4 × 1,250 km = <u>5,000 km</u>.

5. He jogged a total of 7 × 800 m = 5,600 m or 5 km 600 m.

6. 43,500

Puzzle corner. The answer to 14 × 16 × 45 × 50 is 2 × 2 × 5 × 5 = 100 times the answer to 7 × 8 × 9 × 10.

Therefore, the answer is 100 × 5,040 = <u>504,000</u>.

Multiplying in Parts: Another Way, pp. 60-61

Page 60

1. b. 40 × 73 and 8 × 73 c. 10 × 42 and 9 × 42 d. 50 × 89 and 5 × 89

2. a. 20 × 16 = 320; 8 × 16 = 128; 320 + 128 = 448 b. 40 × 73 = 2,920; 8 × 73 = 584; 2,920 + 584 = 3,504
   c. 10 × 42 = 420; 9 × 42 = 378; 420 + 378 = 798 d. 50 × 89 = 4,450; 5 × 89 = 445; 4,450 + 445 = 4,895

Page 61

3. a. 40 × 41 and 6 × 41
   40 × 41 = 1,640; 6 × 41 = 246; 1,640 + 246 = 1,886

   b. 20 × 39 and 8 × 39
   20 × 39 = 780; 8 × 39 = 312; 780 + 312 = 1,092

   c. 10 × 27 + 5 × 27
   10 × 27 = 270; 5 × 27 = 135; 270 + 135 = 405

   d. 90 × 16 + 3 × 16
   90 × 16 = 1,440; 3 × 16 = 48; 1,440 + 48 = 1,488

4. 12 × $27 = $324. To multiply in parts, multiply 10 × $27 = $270, then 2 × $27 = $54, and add those.
   Or, solve it this way: for two months, the bill is $54. For four months, it is $108. Then multiply that times 3 to get $324. Estimates vary. For example, 12 × $30 = $360.

5. 720 Estimates vary. For example, write 1 × 2 × 3 × 4 × 5 × 6 as 24 × 30.
   Estimate that as 20 × 30 = 600, or as 25 × 30 = 750.

6. 12 × 15 kg = 180 kg. To multiply in parts, multiply either 10 × 15 and 2 × 15, OR 10 × 12 and 5 × 12.
   Estimates vary. For example, 10 × 15 kg = 150 kg.

7. 12 × $35 = $420. To multiply in parts, multiply either 10 × 35 and 2 × 35, OR 30 × 12 and 5 × 12.
   Estimates vary. For example, 10 × $35 = $350, or 12 × $40 = $480.

The Standard Multiplication Algorithm with a Two-Digit Multiplier, pp. 62-65

Page 62

1. a. 8 × 65 = 520. 520 + 650 = 1,170
   b. 4 × 82 = 328. 328 + 7,380 = 7,708
   c. 20 × 93 = 1,860. 1,860 + 186 = 2,046
   d. 50 × 70 = 3,500. 3,500 + 210 = 3,710

Page 63

2. a. 636 b. 385 c. 6,396 d. 980 e. 494 f. 950 g. 884 h. 931

3. a. Estimate: 60 × 10 = 600. Answer 728
   b. Estimate: 30 × 60 = 1,800 OR 20 × 70 = 1,400. Answer: 1,625. Here, it helps to round one factor down and one up, since both numbers end in 5. (If you round both up, you get 30 × 70 = 2,100, which is off a lot from the answer).
   c. Estimate: 70 × 40 = 2,800. Answer: 2,860
   d. Estimate: 45 × 10 = 450. Answer: 616
   e. Estimate: 90 × 50 = 4,500. Answer: 4,232
   f. Estimate: 90 × 80 = 7,200. Answer: 7,161

Page 64

4. a. Estimate: 80 × 80 = 6,400. Answer: 6,561
   b. Estimate: 100 × 30 = 3,000. Answer: 3,201
   c. Estimate: 30 × 50 = 1,500. Answer: 1,512

5. a. 15 × 12 = 180 eggs. Estimate: 15 × 10 = 150
   b. 21 × 60 = 1,260 minutes. Estimate: 20 × 60 = 1,200
   c. 11 × 39 = 429. No, 11 buses are not enough. Estimate: 11 × 40 = 440.
   d. 12 × $21 = $252. Estimate: 12 × $20 = $240.

Page 65

6. a. No, it is not. 53 × 61 = 3,233, and 51 × 63 = 3,213. They are close though -- the difference is only 20!

   b. No, the answers are not the same. 42 × 71 = 2,982, and 41 × 72 = 2,952. The difference is 30.

7. The change is $300 − 15 × $17 = $45.

8. She will pay 52 × $98 = $5,096.

9. a. 3,600; 0 b. 300; 12,100

Puzzle Corner:

3 × 105 = 315; 6 × 667 = 4002; 4 × 234 = 936; 5 × 834 = 4,170

Review, pp. 66-68

Page 66

1. a. 1,200; 180
   b. 4,200; 3,300
   c. 81,000; 40,000

2. a. 80; 7 b. 4; 400 c. 300; 80

3. a. 40 b. 2 c. 40

4. In about 8 weeks. 8 × $500 = $4,000.

5. a. Estimation: 7 × 50 = 350; Exact: 336
   b. Estimation: 6 × 800 = 4,800; Exact: 4,878
   c. Estimation: 20 × 20 = 400; Exact: 378
   d. Estimation: 4 × 6,000 = 24,000; Exact: 23,612

Review, cont.

Page 67

6.

Roses	1	2	3	4	5	6	7	8
Price	$0.90	$1.80	$2.70	$3.60	$4.50	$5.40	$6.30	$7.20

7. 2 × 98 = 196; 8 × 17 = 136; 196 − 136 = 60

8. a. 2,000 b. 0 c. 80 d. 20,000

9.

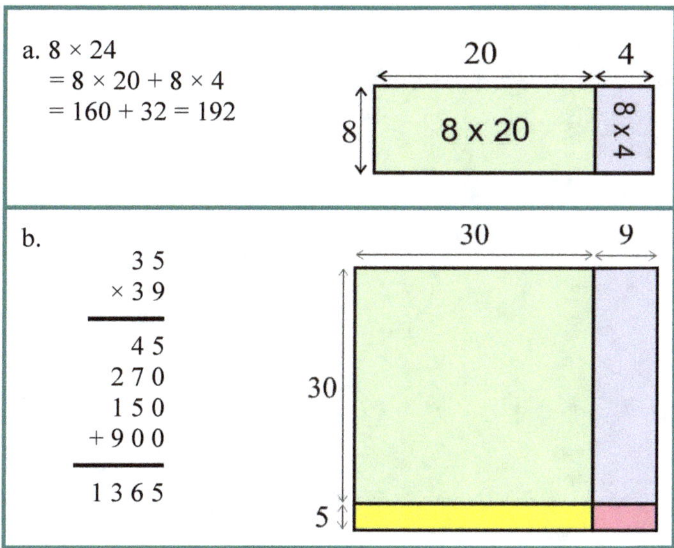

Page 68

10. a. He has 50 × 20 = 1,000 shirts. The cost is 1,000 × $2 = $2,000.
Or, write a single number sentence 50 × 20 × $2 = $2,000.

b. 8 × $2.35 = $18.80; $20 − $18.80 = $1.20. Or, $20 − 8 × $2.35 = $1.20.

c. 5 × $1.50 + $12.50 = $20

d. $45 − $8 = $37. 5 × $37 = $185. Or, 5 × ($45 − $8) = $185.

11. a. Two miles.

Minutes	Miles
5	1
10	2
15	3

b. They would weigh 600 g.

Cans	Weight
1	60 g
7	420 g
10	600 g

More from math MAMMOTH

Math Mammoth has a variety of resources to fit your needs. All are available as economical downloads, and most also as printed copies.

- **Math Mammoth Light Blue Series**
  A complete curriculum for grades 1-7. Each grade level includes two student worktexts (A and B), which contain all the instruction and exercises all in the same book, answer keys, tests, cumulative reviews, and a worksheet maker. International (all metric), Canadian, and South African versions are also available.
  https://www.MathMammoth.com/complete-curriculum
  https://www.MathMammoth.com/international/international
  https://www.MathMammoth.com/canada/
  https://www.MathMammoth.com/south_africa/

- **Math Mammoth Skills Review Workbooks**
  These workbooks are intended to be used alongside the Light Blue series full curriculum, and they provide additional review to the topics studied in the main curriculum, in a spiral manner.
  https://www.MathMammoth.com/skills_review_workbooks/

- **Math Mammoth Blue Series**
  Blue Series books are topical worktexts for grades 1-7, containing both instruction and exercises. The topics cover all elementary mathematics from 1st through 7th grade. These books are not tied to grade levels, and are thus great for filling in gaps.
  https://www.MathMammoth.com/blue-series

- **Make It Real Learning**
  These activity workbooks concentrate on answering the question, "Where is math used in real life?" The series includes various workbooks for grades 3-12.
  https://www.MathMammoth.com/worksheets/mirl/

- **Review Workbooks**
  Workbooks for grades 1-7 that provide a comprehensive review of one grade level of math—for example, for review during school break or summer vacation.
  https://www.MathMammoth.com/review_workbooks/

Free gift!

- Receive over 350 free sample pages and worksheets from my books, plus other freebies:
  https://www.MathMammoth.com/worksheets/free

Lastly...

- Inspire4 is an inspirational website for the whole family I've been privileged to help with:
  https://www.inspire4.com

www.ingramcontent.com/pod-product-compliance
Lightning Source LLC
LaVergne TN
LVHW080327110826
845155LV00026B/212
9781954358713